Contents

Preface		vii
Chapter 1 Introduction		**1**
1.1	Background to the legislation	1
1.2	Adjudication in standard contracts before the Act	4
1.3	What is adjudication?	6
1.4	Appeal and review	11
1.5	Who are the adjudicators?	12
1.6	The future	13
Chapter 2 Construction Contracts and Construction Operations		**16**
2.1	Definition of 'construction operations'	16
2.2	Exclusions from the definition of 'construction operations'	19
2.3	Definition of 'construction contract'	22
2.4	Exclusions from the definition of 'construction contracts'	26
2.5	The mixed contract	30
2.6	Limits of date and place	30
2.7	Residential occupiers	34
2.8	Contracts in writing	35
Chapter 3 The Statutory Right to Refer Disputes to Adjudication		**39**
3.1	Definition of dispute	39
3.2	Required contractual provisions	45
3.3	The incorporation of institutional rules and other terms	55
3.4	The adjudication provisions of the Scheme for Construction Contracts	56

Chapter 4 Starting Adjudication — 57

4.1	Timing	57
4.2	The notice of adjudication	60
4.3	Service of the notice of adjudication	66
4.4	Identification or selection of the adjudicator	67
4.5	Request to an adjudicator nominating body	70
4.6	Terms of agreement with the adjudicator	71
4.7	Procedure if the appointment system fails	73
4.8	Objections to specific adjudicator	75
4.9	Revocation of appointment and resignation of the adjudicator	76

Chapter 5 Preliminary Matters – the Referral Notice and Jurisdiction — 83

5.1	Time for delivery of the referral notice	83
5.2	Form and contents of the referral notice	86
5.3	Related and unrelated disputes	91
5.4	Questions of jurisdiction	93

Chapter 6 Conduct of the Adjudication — 101

6.1	Overriding duties of the adjudicator	101
6.2	The exercise of initiative by the adjudicator	105
6.3	Failure to comply	112
6.4	Representation of the parties	116
6.5	Confidentiality	118
6.6	Timetable for decision	120
6.7	Standard forms of appointment	125

Chapter 7 The Adjudicator's Decision — 127

7.1	The duty to decide	127
7.2	The matters in dispute	129
7.3	Power to open up certificates etc.	134
7.4	Decision on payments	136
7.5	Interest	140
7.6	Form and content of the decision	142
7.7	Binding nature of the decision	150
7.8	Mistakes	152
7.9	Adjudicator's immunity	154

Chapter 8 Costs — 158

- 8.1 Adjudicator's right to fees and the power to apportion — 158
- 8.2 Right to require security for his fees — 165
- 8.3 Power to order payment of costs — 169

Chapter 9 Enforcement — 174

- 9.1 The Act and the Scheme — 174
- 9.2 Application for summary judgment — 178
- 9.3 Other enforcement procedures — 181
- 9.4 Challenges to enforcement — 185

Chapter 10 Payment — 205

- 10.1 Introduction — 205
- 10.2 The right to stage payments — 206
- 10.3 Timing and quantification of payments — 208
- 10.4 Notice of amount to be paid — 209
- 10.5 Notice of intention to withhold payment — 210
- 10.6 Right to suspend — 212
- 10.7 Conditional payment provisions — 214
- 10.8 The Scheme — 218

Appendix 1 Housing Grants, Construction and Regeneration Act 1996 — 232
Appendix 2 The Scheme for Construction Contracts (England and Wales) Regulations 1998 — 241
Table of Cases — 253
References to Housing Grants, Construction and Regeneration Act 1996 — 257
References to Scheme for Construction Contracts — 258
Subject Index — 259

Preface

The first four months of 1998 were a period of frantic activity for construction lawyers. We had all been astonished at the way in which Sir Michael Latham's suggestions for the regulation of payment terms in the construction industry had been adopted by a Conservative Government, who most had thought would have an antipathy to such legislation. We were frankly terrified by the prospect of having to deal with the new dispute resolution system that somehow would produce decisions in just 28 days. We had spent the previous two years warning our clients that something quite remarkable was about to happen, and now the days were counting down. Contracts were frantically being drafted and revised, main contractors were asking us to devise complex avoidance measures and some of us were spending our weekends attending courses to train to be adjudicators.

Not everyone wanted adjudication to work, and many of its supporters feared that it would fail. Several highly respected and very senior construction lawyers argued that it was offensive to the traditions of the common law and would lead to serious injustice. Many thought that the courts would find it difficult to enforce the decisions of adjudicators and for this reason the system would never get off the ground.

So it was that despite enormous excitement, there was a great deal of nervousness as the construction industry awaited the new dawn on 1 May 1998, the day when Part 2 of the Housing Grants, Construction and Regeneration Act 1996 was to come into force. Like so many dawns in these rather damp islands, this one was something of a disappointment. There was no blinding flash of transformation for the construction industry. Weeks went by with no apparent changes at all. Standard forms of contract had been changed, but many old editions were still being used. New instructions for the lawyers all seemed to concern contracts made months before, and adjudication was not available.

But something in the undergrowth was stirring, and by February 1999 one adjudication decision found its way to the Technology and Construction Court. It was enthusiastically greeted by Mr Justice

Dyson. Others followed in a steadily accelerating flow, most with a similar success, and it became clear that adjudication was working. As a result more and more construction businesses are taking their disputes to adjudication. It is not just the weapon of the oppressed subcontractor; many employers and main contractors are using the procedure to achieve a rapid and comparatively inexpensive resolution of a seemingly intractable argument.

Several excellent guides to this new process were published shortly before or shortly after the Act came into force in May 1998. I have enormous admiration for their authors, who were obliged to speculate about how and indeed whether adjudication would work. Two years later, with a host of court decisions on which to draw, I have had a much easier task. While there will doubtless be a continuing stream of new cases in which adjudicator's decisions in particular circumstances will be tested, the principles are now established. There is certainly no longer any need to speculate about whether or not adjudication will succeed. With disputes being referred to adjudicators at an estimated rate of 2000 per annum there is no longer any doubt.

The speculation now is on how far the process will spread. Why should it be confined to the construction industries? The technology and construction industries are served by the same specialist court which has heartily endorsed adjudication. When will the technology industry demand the same benefit as construction? How about transport? In twenty years' time the advent of adjudication may be seen as a far more significant development in dispute resolution than the reforms of civil procedure with which lawyers have also been grappling in recent times.

I apologise to my female friends and colleagues who may be offended by my consistent reference to the adjudicator as 'he'. I have no doubt that the day is coming when my computer will tell me how to move randomly between gender so as to achieve complete equality, but until then it seems simpler to stick to 'he' and 'him' and ask the reader to interpret appropriately. The Housing Grants, Construction and Regeneration Act 1996 is referred to as 'the Act' unless the context is likely to give rise to confusion, and I have shortened the Scheme for Construction Contracts to 'the Scheme'.

I started writing this book while a partner at Laytons and completed it as a partner at Osborne Clarke OWA. In both firms I have had enormous support, and many a happy hour debating finer points of the adjudication process. Many of the better ideas have come from my colleagues, but of course any mistakes are my own.

I must also thank the publishers Blackwell Science and in particular Julia Burden whose enthusiastic encouragement has been invaluable. Finally, I should also remember my son Sam, who I have had to push off our computer rather too often in recent months.

I have aimed to state the law as it stands at 1 November 2000.

John Redmond
Osborne Clarke OWA
Bristol

CHAPTER ONE
INTRODUCTION

1.1 Background to the legislation

The Housing Grants, Construction and Regeneration Act 1996 came into force on 1 May 1998. This book is about Part II of that Act, which itself is often known as the 'Construction Act'. That single statute is so fundamental to everything discussed in this book that, with the exception of one or two passages where there might be some confusion, it is generally referred to in this book as 'the Act'. Housing Grants and Regeneration, each important pieces of legislation in their own right, have nothing to do with the remarkable Part II. The most significant piece of legislation to affect the construction industries within the lifetime of anyone currently working within them was hidden between two parts of the Act that for most will remain unread.

It is natural to ask where the story started that led to this extraordinary statute. The obvious answer is the report *Constructing the Team* published in July 1994. On 5 July 1993 the House of Commons was told that there was to be a 'Joint Review of Procurement and Contractual Arrangements in the United Kingdom Construction Industry'. The result was not a Government report, prepared by a commission striving to find consensus and achieving compromise, but a wholly personal report by one man – Sir Michael Latham. His task was to define and address problems perceived but insufficiently defined within the construction process. Latham summarised his purpose in the foreword to the report:

> 'The Review has been about helping clients to obtain the high quality projects to which they aspire.'

Latham made 30 principal recommendations, many of which had several subsidiary suggestions. They covered an enormous spectrum of issues, including aspects of public procurement, training and professional education, quality control, technical research and latent defects insurance. None of these found their way into the Act.

There were however several recommendations that can be seen as direct precursors to the Act that was to follow four years later:

(1) There should be a system of standard form contracts usable in all construction and similar projects, covering all aspects of the project from appointment of the first design consultant to the engagement of the last subcontractor. That system should be based on the New Engineering Contract (also known as the Engineering and Construction Contract), with some amendment to embrace 13 requirements set out by Latham. One of those requirements was provision for speedy dispute resolution by an impartial adjudicator, referee or expert.
(2) Both public and private sector clients should use the New Engineering Contract.
(3) Legislation should be introduced prohibiting the amendment of the standard form with regard to payment and interest, reinforcing the right to refer disputes to adjudication, requiring advance notice of set-off, and protecting similar rights in bespoke contracts.
(4) Adjudication should be the normal method of dispute resolution.

Dispute resolution in the construction industry was a particular concern, and not only for Sir Michael Latham. While he was reviewing contract practices in the industry, Lord Woolf was carrying out a review of the English court system. He was considering the problems of time and cost in litigation. Annex 3 to the 1996 Woolf Report gave a fascinating insight into these aspects of actions in the Official Referees' courts, now renamed the Technology and Construction Court.

In 205 such actions, normally involving construction, the mean duration from instruction to conclusion was 34 months, with the median at 30 months. Costs as a percentage of claim value averaged 158% in claims of less than £12,500 and 96% in claims of £12,500–£25,000. The costs were not the actual costs incurred, but the costs that the losing party had been ordered to contribute to the winner through the process of taxation, or court assessment. On the basis that taxed costs probably represented 75% of actual costs paid by the party, the total figure would be 30% higher, and could then be doubled to account for one more party. Thus a typical claim for £10,000 would involve costs of £41,000. Working through the figures provided in the annex, the costs of a two party action involving a typical £200,000 dispute would total at least £165,000.

Official Referee cases were consistently the most expensive of those reviewed.

These statistics came as no surprise to practitioners in construction litigation. Lawyers were well aware of the high level of costs and the time that was taken to bring a case to a conclusion, both in court and in arbitration. Indeed many lawyers felt that they were giving a fast and economic service. In comparison with several other jurisdictions in Western Europe, they had some reason for this complacency. But such a comparison was of little comfort to an industry that saw too much of its meagre profitability being drained in dispute resolution, often long after the completion of the relevant building projects.

There was widespread agreement with Latham's conclusions that the industry needed an improved method of dispute resolution, but it was not immediately obvious what form that improved method should take. Latham inferred that the adjudication provisions of the New Engineering Contract were largely satisfactory, and the increased or possible mandatory use of that contract would lead to adjudication on a more widespread basis.

Discussion about Latham's proposals concentrated on other issues, particularly with regard to payment. He had suggested the establishment of a trust fund that would ensure payment of all contractors and subcontractors in the event of an insolvency in the chain of contracting parties, and issues such as this attracted rather more attention and controversy. Notwithstanding the enormous energy and enthusiasm with which Sir Michael Latham presented and supported his proposal, most commentators expressed some doubt as to the likelihood of sufficient consensus developing to enable any substantial reform to be introduced with statutory backing.

Anyone listening to the Queen's Speech at the opening of the new parliamentary year in November 1995 would have been unlikely to form the view that significant legislation affecting construction contracts was to be expected within the coming months, but the additional material contained in Government press releases about the legislative programme at the time did contain a reference to the intention to introduce a Bill. The following February, the Housing Grants, Construction and Regeneration Bill was introduced in the House of Lords. After vigorous debate in both Houses of Parliament, and despite a general election that brought about a change of Government, the Bill completed its passage and became the Act in the same year.

There then followed a period of consultation leading to the preparation of the draft Scheme for Construction Contracts. This was not an easy task. A draft had been prepared during the passage of the bill through Parliament, and had proved so controversial that the legislation was close to being abandoned. That draft had been wholly misconceived, for example providing that an adjudicator's decision would be final, without any subsequent opportunity for reconsideration. Fortunately it was made clear in debate that the Scheme would not be rushed and that no assumptions should be made about what it would contain on the basis of the first draft.

The consultation process was based on a document entitled 'Making the Scheme for Construction Contracts.' Once again there were several aspects of the process that were to change before the final version was published and eventually passed into law by the making of the Scheme for Construction Contracts (England and Wales) Regulations in March 1998. Misconceptions remain about the Scheme based on the contents of the consultation document, and care still has to be taken to ensure that reference is made to the Scheme as it is, rather than as it was anticipated (for example with regard to the effect of failure to give notice of the sum that is to be paid, discussed in Chapter 10).

1.2 Adjudication in standard contracts before the Act

The word 'adjudication' is not new to the world of construction contracts. It made its first appearance in 1976 when the standard form of nominated subcontract for use in the JCT system (the 'Green Form') was amended following the House of Lords decision in *Gilbert Ash (Northern) Ltd* v. *Modern Engineering (Bristol) Ltd* (1974). Prior to that case, it had been understood an architect's certificate gave rise to a debt of an unusual kind. It had some of the characteristics of a cheque, and it was not possible in normal circumstances for the party with the obligation to pay to exercise a right of set-off. The House of Lords found that this was a fiction. The obligation to pay under a construction contract, whether or not supported by a certificate, was no different to any other debt. In the absence of specific agreement in the contract, there was no reason why normal common law rights of set-off could not be exercised.

In order to protect subcontractors against unjustified exercise of set-off rights by main contractors, the Green Form was amended to introduce conditions that had to be satisfied before set-off could be taken, and a system of adjudication was provided to resolve disputes about such matters. This amendment was also introduced to the Blue Form, for domestic subcontractors. The system was carried through to the new forms of subcontract NSC/C and DOM/1 used with the new form of main contract JCT 80, and the other JCT family of contracts.

In summary, the main contractor was entitled to deduct any sum agreed to be due to it from the subcontractor, or found to be due in arbitration or litigation. If those circumstances did not apply, the main contractor would only be able to make a deduction if he gave written notice of his intention to do so, with detailed and accurate quantification at least three days before the payment to the subcontractor became due for payment.

If the subcontractor disputed the intended set-off, it could give notice of that dispute and could raise a counterclaim. It could only do so, however, within 14 days of receipt of the main contractor's notice. At the same time the subcontractor was obliged to serve notice of arbitration and also request action by the adjudicator, who would normally have been named in the subcontract. The contractor then had 14 days in which to serve a defence to any counterclaim that had been raised. Seven days later, the adjudicator was required to produce his decision on the matter, which was binding until later arbitration or settlement. He would make his decision entirely on the basis of the written submissions, and would make no further enquiry.

This adjudication procedure was well known, but not often used. It was strictly limited to disputes about set-off by main contractors and could only be used if the subcontractor acted within the 14 day period. As will be seen when the Act's system of adjudication is considered in later chapters, it was quite different to the comprehensive process that was to be introduced in 1998.

The JCT forms of main contract first experimented with adjudication in 1988, when the Standard Form of Contract With Contractor's Design (JCT 81) was amended so as to introduce Supplementary Provisions.

The Supplementary Provisions, and in particular Supplementary Provision S1 dealing with adjudication, were not popular and were often deleted. If S1 was left in, however, any dispute arising prior to practical completion on any of a wide variety of matters had to be submitted to adjudication rather than arbitration, although it could be referred on later to arbitration if the dispute continued. It was intended that the process should be speedy, but there was no time limit for the making of the decision. Once practical completion had been reached, the process was not applicable.

This mechanism was deleted from the standard form in 1998

when the Act came into force. In the ten years of its existence, very few adjudications had taken place, and there had been no enthusiasm to extend the application of the Supplementary Provision to the other JCT contracts.

Sir Michael Latham had been particularly enthusiastic about the New Engineering Contract. That contract had included adjudication as the mandatory dispute resolution system. But even that adjudication process bore little resemblance to adjudication as it was to appear in the Act. In common with other aspects of procedure under this form of contract, adjudication was subject to strict time limits. If the dispute was about an action or inaction of the project manager or the supervisor, the contractor had just four weeks from becoming aware of the problem to notify the dispute to the project manager. Once it had been notified, the contractor was able to submit the dispute to the adjudicator, but had to wait two weeks before doing so, and had just two further weeks in which to make the submission. Far from being able to refer a dispute to adjudication *at any time*, as was to be provided in the Act, there was just a two week window in which the contractor could initiate the process. Similar time limits applied to a claim being made by the employer.

Once the process was started, there was to be a four week period for submission of information by everyone involved, and then a further four week period for the adjudicator to make his decision. These periods could be extended by agreement. Even without extension, the process was to take twice as long as adjudication under the Act.

1.3   What is adjudication?

The account of some of the previous systems of adjudication found in construction contracts demonstrates that even within one industry there is no one procedure that is entitled to call itself 'adjudication'. Once we stray outside the world of construction contracts we are faced in addition with adjudication within the realms of immigration law or VAT, which bear little resemblance to the process described in this book. The question 'What is adjudication?' is therefore meaningless unless it is qualified by adding a reference to the particular model of adjudication that is being discussed. In this book we are dealing with adjudication as it is found in the construction industries as a result of the requirements of the Act.

The obvious place to start in a search for a definition of this creature is the statute that created it. We search in vain. There is no definition within the Act, nor is there one within the Scheme. We can retreat into tautology by defining adjudication as a system of dispute resolution that complies with the requirements of the Act. This may be strictly correct, but it is hardly informative.

The desire to find a neat definition is particularly strong when a litigation lawyer from another jurisdiction asks for a simple explanation of the process. It is tempting to describe it by reference to some other form of dispute resolution process, but whichever process is chosen, it is clear that the differences are more obvious than the similarities. It may be helpful though to work through this process in an attempt to develop a definition of adjudication by demonstrating what it is not.

1.3.1 Litigation

The obvious reason that this comparison can be dismissed is that litigation involves courts, deriving their authority from the law of the land. Courts do not need consent to their jurisdiction through contractual terms. Citizens, human or corporate, rely on the law to give them rights and they rely on the courts to enforce their rights. They submit to the jurisdiction of the courts so that they can claim the protection of the courts when required. Adjudication however requires a contract. If there is no contract in existence between the parties there can be no adjudication between them either. Adjudication is only available because there is a term of that contract that disputes can be referred to the process. The Act may require construction contracts to contain such a term and to that extent contractual freedom may be fictional, but nevertheless adjudication has to rely on that fiction.

Litigation involves resolution by a judge, who is a servant of the state, and the judge owes his duties to the state rather than to the parties who appear before him. The adjudicator is appointed by the parties and is paid by them. He owes a duty to no one other than the parties themselves. Moreover he can be dismissed by them if he fails to carry out his duty.

The result of litigation is immediately enforceable by agencies of the state. Typically these include execution on the assets of the losing party, insolvency procedures or even in some cases imprisonment. Adjudicators' decisions have been supported vigorously by the courts, but enforcement is not possible without further consideration by the courts, and conversion into court judgments.

Litigation is the last resort. After the courts, there is nowhere else for the parties to go. There may be an appeal system available from a court, but that appeal will be to another court, and finally the ultimate appeal tribunal is reached. An adjudicator's decision is only binding on the parties until the matter has been decided by arbitration, litigation or agreement.

It would however be wrong to say that there is nothing in common between adjudication and litigation. The Act does not say that the adjudicator is to arrive at his decision in accordance with the law, but that is inferred from the requirement that the adjudicator is to be able to take the initiative in ascertaining the facts and the law (section 108(2)). It would seem therefore that the law must be of some relevance, although the decision will not be invalid if incorrect factually or legally. Furthermore the Act says nothing about any requirement to act in accordance with natural justice, but at least one decision in the Technology and Construction Court (*Discain Project Services Ltd* v. *Opecprime Development Ltd*, August 2000) suggests that respect for the rules of natural justice is required. This case is discussed further in Chapter 9.

1.3.2 Arbitration

There are more obvious similarities between adjudication and arbitration than between adjudication and litigation. Many adjudicators are also arbitrators, and vice versa. Both are private systems, and both rely on contracts to give them authority. Both offer the tribunal, whether adjudicator or arbitrator, the facility of exercising his initiative to ascertain the facts and the law. Both require the operation of the courts to enforce their conclusions, but there are significant differences.

Like judgments of the courts, arbitration awards are normally final and conclusive. By agreeing to submit disputes to arbitration, the parties agree to replace the public court system with a private process. They do that for a variety of reasons. They may wish to keep their disputes confidential; they may prefer to have an expert in their particular trade or profession decide their disputes; or they may believe that arbitration is a more flexible, speedy or economic procedure than litigation. In any event, arbitration is not seen merely as a first stage in dispute resolution, providing a temporary balance whilst full arguments are prepared for trial elsewhere.

Unless there is serious dissatisfaction with the result of the arbitration, and grounds for appeal or application to court on the basis of some irregularity in the process, the arbitrator's award is the final determination of the problem. Adjudication is not necessarily final at all, as explained above.

Arbitration is governed by the comprehensive framework provided by the Arbitration Act 1996. The Housing Grants, Construction and Regeneration Act does not provide anything comparable. Moreover, whereas the Arbitration Act may set out what will happen if the parties choose to make an arbitration agreement between themselves, there is absolutely no compulsion on anyone to include an arbitration agreement in any of their business dealings. The adjudication legislation on the other hand requires adjudication to be available in virtually every construction contract whether the parties want it or not. Unlike arbitration though there is no obligation on the parties to avail themselves of adjudication even if they have voluntarily incorporated it in their contract. If a claimant party ignores his right to refer a dispute to adjudication and issues court proceedings, those proceedings will not be stayed in favour of adjudication.

1.3.3 Expert determination

In *Bouygues UK Ltd* v. *Dahl-Jensen UK Ltd* (November 1999) Sir John Dyson drew an analogy between adjudication and expert determination. He applied a test that had developed in expert cases, *Jones* v. *Sherwood Computer Services plc* (1989) and *Nikko Hotels (UK) Ltd* v. *MEPC plc* (1991). In *Nikko Hotels* Mr Justice Knox had said this:

> '... the expert's decision will be final and conclusive, and, therefore not open to review or treatment by the courts as a nullity on the ground that the expert's decision on construction was erroneous in law, unless it can be shown that the expert has not performed the task assigned to him. If he has answered the right question in the wrong way, his decision will be binding. If he has answered the wrong question, his decision will be a nullity.'

This was the express basis of Sir John Dyson's approach when asked to enforce an adjudicator's decision, and that approach was enthusiastically approved by the Court of Appeal.

This does not mean that adjudication is similar to the process of expert determination. Once again a principal difference is the fact that the adjudicator's decision is not the final result unless the parties want it to be, but there are other distinctions. The expert is employed to use his expertise. If the dispute is about the value of a shareholding in a business, or the market rent of property, the expert will be expected to be able to form a view on the basis of his knowledge and experience. An adjudicator will often have little direct experience on which to base his decision, being perhaps a lawyer appointed to decide a dispute about the valuation of a variation, or an engineer appointed to decide a point of law. He is given power to exercise his initiative in ascertaining the facts and the law, but he is expected to make his decision on the basis of what he ascertains.

1.3.4 Mediation

The relatively informal procedures adopted by many adjudicators cannot be confused with the various processes commonly called 'alternative dispute resolution'. Such processes are entirely voluntary, and their aim is to resolve the dispute by compromise, encouraged by a facilitator. The resolution, if achieved, will be binding on the parties because they want it to be binding, and it will then be enforceable if necessary by action based on contract. The process will not normally produce a result that is binding only until further arbitration or litigation. Mediation is a structured negotiation, but there is no place for negotiation in adjudication. The adjudication procedures may encourage negotiation between the parties, and it is even possible that the adjudicator will also be appointed to act as mediator, but it is not the adjudicator's job whilst acting as adjudicator to mediate or facilitate negotiation.

Having established that adjudication is not litigation, arbitration, expert determination or mediation, it becomes clear that in adjudication we have an entirely new dispute resolution system. It has little in common with any of the conventional systems with which we have worked before. It is dangerous to assume that rules that are taken for granted in some other system are applicable to adjudication. It is not right to assume that because an arbitrator can or cannot do something as arbitrator, he will or will not be able to do something similar in adjudication. Analogies may be drawn, but they must be drawn with care. It is therefore not surprising that after a relatively slow start in 1998, with no cases being reported in the Technology and Construction Court until February 1999 (*Macob Civil Engineering Ltd* v. *Morrison Construction Ltd*), the flow of reported judgments rapidly increased and by late 2000 had reached a flood, with several such reports per month. Each case that is reported adds a little more to our understanding of how adjudication should work, and indeed how it does work in practice. Basic principles are becoming established, and are described throughout the chapters that follow, but it will be several years before the law and practice of adjudication will be fully understood, and only then will it be possible to answer the question 'What is adjudication?' in a comprehensive authoritative and meaningful way.

1.4 Appeal and review

Decisions of adjudicators are not the end of the matter. Section 3 of the Act provides that the decisions are to be binding, but only until finally determined by arbitration, litigation or settlement. Whatever the outcome of the adjudication, the matter can be raised afresh in full legal proceedings, or, if there is an arbitration agreement, in arbitration. If one or other of the parties wishes to take the matter on to litigation or arbitration, either on its own or as part of a wider dispute, the adjudication will be of no significance in those proceedings. The judge or arbitrator will approach the dispute without reference to the adjudicator's decision. He may be made aware of the fact that an adjudicator made a decision on one or more of the points in the case, but he will not start from that decision and decide whether it was right or wrong. The whole decision-making process will begin anew.

The final determination in litigation or arbitration may be quite different from the adjudicator's decision, but there is no right of appeal from the adjudicator. An adjudicator's decision will be enforced even if it is patently wrong. The only effective challenges to enforcement have been to demonstrate that the adjudicator acted outside his jurisdiction or behaved in serious breach of the principles of natural justice with regard to a significant matter or that the receiving party was insolvent. These decisions are discussed in detail in Chapter 9.

It has been suggested, particularly by Anthony Speaight QC in a paper delivered at a King's College Conference in July 2000, that adjudicators' decisions are in some instances subject to judicial review. This would only apply to adjudication proceedings in connection with contracts that did not contain any agreed system of adjudication complying with the requirements of the Act. In such contracts the Act applies the Scheme, and thus a statutory procedure is imposed on the parties. The adjudicator's authority in such an adjudication is derived from the Act, and not merely from private agreement of the parties, although the Act works by implying terms into the contract. That analysis would suggest that the adjudicator is exercising a public law function, and that if the decision is wrong in law or irrational, or the adjudicator acted improperly, the decision could be overturned in an application for judicial review. This theory has yet to be put to the test.

1.5 Who are the adjudicators?

The Act does not give any guidance as to who should be appointed as adjudicator, how the adjudicator should be selected, or what qualifications the adjudicator should have. As we will see in Chapter 4, the appointment will be governed by any express terms of the contract, but in default of agreement it may be made by 'an adjudicator nominating body'. There is no statutorily defined or controlled list of such bodies, and no requirement that an adjudicator nominating body should have any professional standing. Any organisation can offer to perform the task. Similarly there is no requirement that the adjudicator should be professionally qualified or have any relevant experience. There is no reason under the Act or the Scheme why a local parent teachers association should not set itself up as an adjudicator nominating body. When requested to appoint an adjudicator it could quite validly select the teacher in charge of physical education, with the advantage of experience in refereeing football matches. His decision would be as valid and enforceable as any other adjudicator's decision.

In practice, of course, organisations with no interest in the construction industry are not involved. In December 1999 a survey by Glasgow Caledonian University Department of Building and Surveying reported that there were 21 bodies currently offering their services in appointing adjudicators. As there is no register of such bodies it is impossible to state with certainty how many bodies are active at any one time. The Construction Industry Council (CIC) has undertaken the task of co-ordinating the nominating bodies, and reported in July 2000 that among the 14 bodies with whom they communicated, there were 957 listed adjudicators. As many adjudicators are on more than one list, it was suggested that the total number of individuals then offering to act as adjudicators was about 400.

Most if not all of these organisations require candidates for inclusion on their lists to undergo professional training in the skills of adjudication before being admitted. Some, such as the Chartered Institute of Arbitrators (CIArb), run their own training courses, whilst others accept candidates who have trained with another adjudicator nominating body. Some, but not all, require a specific length of experience in dealing with the construction industry. Most include within their lists professions other than that principally represented by the organisation. The Royal Institution of Chartered Surveyors (RICS), for example, has solicitors on its list and the Technology and Construction Solicitors Association (TeCSA) includes surveyors and engineers.

Adjudication is growing in popularity. The nominating body that has been called upon most often to provide an adjudicator is the RICS. In 1998 the RICS was called upon to appoint 23 adjudicators. In 1999 the figure rose to 377. In March 2000 the RICS reported that requests to appoint were being received at a rate of 75 per month, and by November 2000 the rate had increased to 100 per month.

As may be expected, many adjudicators are qualified as quantity surveyors; 57% of those surveyed in research by the University of Wolverhampton published in July 2000 were quantity surveyors, whereas only 4% were lawyers. Engineers accounted for 26% and architects for 23%. Several were dual qualified.

1.6 The future

Adjudication appears to be popular, in that the rate of appointments is accelerating. It was not however universally welcomed. The whole process was viewed with intense suspicion by many distinguished lawyers during the passage of the litigation and in the period immediately before the Act was brought into force. Their concerns were both theoretical and practical. The adjudication process appeared to usurp the position of the courts and traditional arbitration in a way that was likely to prove offensive to natural justice. Furthermore it was an unjustified interference with the rights of commercial parties to agree their terms of business.

Turning to practical issues, the extraordinarily short time period in which the adjudication was to be conducted meant that there were bound to be errors, which would lead to serious injustice. Finally, many decisions would be unenforceable because of the inability of the courts to act when a contract contains an agreement to refer disputes to arbitration.

These concerns have not finally been allayed, but the strength of feeling on such issues is fading. As will be seen in later chapters, the courts, both of first instance and appeal, have been extremely supportive. The judges of the Technology and Construction Court have not merely acquiesced, allowing their procedures to be used to enforce, but they have adapted the procedures of the court so as to match the speed of the adjudication process. Effective enforcement is certainly not the problem that it was anticipated to be. Adjudication has not usurped the power of the courts but given the courts a new relevance to the construction industry.

Questions about natural justice, coupled with issues raised by the Human Rights Act, still arise, but the courts appear to be addressing them in a way that will enable adjudication to continue its work. Indeed consideration of these questions is leading to a better understanding of the adjudication process and the role that it plays. There have been clear examples of errors in decisions, but the temporary nature of adjudication enables such errors to be rectified.

Academic lawyers were not the only ones to view adjudication with concern. Whilst subcontractors warmly welcomed adjudication, especially as it came attached to the Act's requirements regarding payment, many main contractors saw the legislation as a considerable threat. They attempted to devise avoidance techniques that would enable them to continue with their previous payment practices. Many, although not all, such techniques have failed, and the energies of most main contractors appear to have been diverted to finding ways of making adjudication work for them rather than against them. Adjudicators are increasingly being asked by parties who traditionally would have been defending claims, to give negative decisions to the effect that claims do not exist.

Insurance companies were particularly worried about the extent of the new system. Whereas it had been understood that adjudication would affect contractual disputes about time and money, consultants found that they were also involved. Moreover there is no reason why a client cannot bring to adjudication a claim against his architect several years after the completion of the building work when a defect appears. Insurers have previously had the option of pushing the client to expensive and prolonged litigation, which on occasions could be a valuable tool in negotiation. Now however the insured's liability might be established, albeit on a temporary but enforceable basis, in just 28 days.

To meet this risk, some insurers introduced standard endorsements on policies that made it quite likely that the insured would have no effective cover. Typical conditions precedent for cover included a requirement that the adjudication provisions in a contract would be no more onerous to the insured than those contained in the Scheme, although how this was to be judged is difficult to speculate. Further, such endorsement included a condition precedent that any notice of intention to adjudicate would be notified to the insurer within two working days of receipt. Many professional practices would find it difficult to say with absolute confidence that it was practical to guarantee compliance with such a requirement.

The involvement of insurers in the adjudication process is still very much in its infancy, as clearly relatively few substantial projects based on contracts dated after 1 May 1998 have yet developed disputes about defects. This may be an area of substantial conflict in the future.

General opinion in the construction industry and in the professions that serve it appears to be favourable to adjudication. Satisfaction is not universal, and there are calls for aspects of the procedure to be tidied up or improved. In particular there are calls to deal with the following:

- Costs – the ability of the adjudicator to award costs between the parties
- Security for the adjudicator's fees
- Consistency in the fees charged by adjudicators
- The correction of errors by the adjudicator
- The ability of the adjudicator to deal with more than one dispute
- Clarification of the adjudicator's ability to award interest

Whilst many would agree that these matters are in need of clarification or review, there are many different opinions about how they should be clarified or reviewed.

Nevertheless the accelerating use of adjudication suggests that the process is perceived to be working. There does not appear to be any public expression of the opinion that adjudication has failed and that it should be abandoned in favour of a return to the dispute resolution regime that existed before 1 May 1998. It seems that adjudication as created by the Housing Grants, Construction and Regeneration Act 1996 is here to stay.

CHAPTER TWO
CONSTRUCTION CONTRACTS AND CONSTRUCTION OPERATIONS

As explained in the previous chapter, adjudication is a word that has been used to describe various dispute resolution processes in construction projects for many years. This book is about only such processes created by Part II of the Housing Grants, Construction and Regeneration Act 1996. The Act introduced a method of dispute resolution that is quite different to any of its namesake predecessors. Being a creation of the statute, however, it only applies to such contracts as are brought within its operation by the Act. It does not apply to any other contract unless the parties agree to refer the dispute to the statutory adjudication process as if it were within the Act's scope.

If the contract is to be caught by the Act, it must be a 'construction contract'. In order to be a construction contract, it must involve or relate to 'construction operations'.

2.1 Definition of 'construction operations'

The first question must therefore be whether the work to be done is within the definition of construction operations, or relates to such work. Construction operations are defined in section 105(1) of the Act:

> '105–(1) In this Part "construction operations" means, subject as follows, operations of any of the following descriptions –
> (a) construction, alteration, repair, maintenance, extension, demolition or dismantling of buildings, or structures forming, or to form, part of the land (whether permanent or not);
> (b) construction, alteration, repair, maintenance, extension, demolition or dismantling of any works forming, or to form, part of the land (without prejudice to the fore-

going), walls, roadways, power-lines, telecommunications apparatus, aircraft runways, docks and harbours, railways, inland waterways, pipelines, reservoirs, watermains, wells, sewers, industrial plant and installations for purposes of land drainage, coast protection or defence;
(c) installation in any building or structure of fittings forming part of the land, including (without prejudice to the foregoing) systems of heating, lighting, air-conditioning, ventilation, power supply, drainage, sanitation, water supply or fire protection, or security or communications systems;
(d) external or internal cleaning of buildings and structures, so far as carried out in the course of their construction, alteration, repair, extension or restoration;
(e) operations which form an integral part of, or are preparatory to, or are for rendering complete, such operations as are previously described in this subsection, including site clearance, earth-moving, excavation, tunnelling and boring, laying of foundations, erection, maintenance or dismantling of scaffolding, site restoration, landscaping and the provision of roadways and other access works;
(f) painting or decorating the internal or external surfaces of any building or structure'

This list may seem a remarkable attempt to achieve the impossible task of defining one of the economy's most diverse industries, but it is not entirely original. A similar definition, but not exactly the same, is to be found in the Income and Corporation Taxes Act 1988, section 567.

The use of the word 'building' does not add to the meaning of the word 'structure':

' "Structure" is anything which is constructed, and it involves the notion of something which is put together, consisting of a number of different things which are so put together or built together, constructed as to make one whole, which is then called a structure.'

(Mr Justice Humphreys in *Hobday* v. *Nash* (1944))

'Every building is a "structure".'

(*Mills & Rockleys Ltd* v. *Leicester City Council* (1946))

The activities described in subsection (a) are self-explanatory, save that the construction etc. of structures is only included if those structures form or are to form 'part of the land'. This phrase relates to the principle that goods that are fixed to the land become the property of the owner of the freehold of that land, subject to the interests of any relevant leaseholder or other person with an interest in the land (*Minshall* v. *Lloyd* (1837)). A 'Dutch barn' standing on sockets dug into the ground has been found not to be a fixture and therefore does not form part of the land (*Culling* v. *Tufnal* (1694)). Hence the deposit on a site of a portable cabin which is not fixed in any way other than by gravity would not fall within this definition.

It should be noted that maintenance is included within the definition. Facilities management services, undertaken by contractors who may not have considered themselves to be within the mainstream of the construction industry, may well include construction operations.

Subsection (b) extends the definition to works, again subject to the 'forming part of the land' qualification. A number of civil engineering operations are listed, but these are not exclusive. It was held in *Palmers Ltd* v. *ABB Power Construction Ltd* (Judge Thornton, October 1999) that the work of fabrication or erection of an item of plant on a construction site but not at its final position, so that it can be subsequently moved into place, is within this description.

Subsection (c) extends the definition further, this time by including the mechanical and electrical engineering and associated sectors of the construction industries. Once again, it goes further than the conventional concept of construction. Installers of security systems, for example, have been surprised to find that their activities are now construction operations. The 'forming part of the land' limitation again applies, however, and suppliers of freestanding heaters, lights, air-conditioning equipment and fire extinguishers are therefore not engaged in construction operations. Curiously this part of the definition does not include alteration, repair, maintenance, extension, demolition or dismantling. This led to an argument by a property owner that the maintenance of heating systems was not a construction operation and not covered by the Act. Mr Justice Dyson, in *Nottingham Community Housing Association* v. *Powerminster Ltd* (June 2000) rejected this. The maintenance of heating systems clearly fell within subsection (a) and was therefore a construction operation. The fact that subsection (c) was expressed in more limited terms did not affect the clear meaning of the earlier subsection.

Subsection (d) covers cleaning carried out in the course of construction etc., but not otherwise. This may produce some difficulties. It might be argued that regular cleaning of parts of the building constitutes maintenance, in which case it is covered by subsection (a). The cleaning of the exterior of the building may be restoration, in which case it is covered, or it may not. It is clear that cleaning of the site before practical completion is a construction operation, as is the regular cleaning of the carriageway during road surfacing works.

Subsection (e) covers a miscellany of matters that are incidental to the activities described in the previous subsections. They do not need to be undertaken at the same time as the other activities. The laying of foundations for example might be carried out some considerable time before the construction of the building on those foundations. Arguably the work in the foundations is caught in its own right in subsections (a) or (b), but in case it escapes it is included in (e).

Scaffolding is expressly mentioned. Other temporary works such as falsework and formwork will also be caught here although they are not specifically included. Landscaping is in the list in subsection (e) only if it is incidental to other operations, but once again it might be argued that landscaping works, undertaken for their own sake, are in reality within subsection (b). In *Palmers Ltd* v. *ABB Power Construction Ltd* it was held that scaffolding that was preparatory to the construction of a major item of plant, itself within section 105 (1)(b), was caught by section 105(1)(e), even though the construction of the plant item was excluded by section 105(2). This case is discussed further below.

Finally, subsection (f) includes all painting or decorating of internal or external surfaces of any building or structure. This suggests that painting of pipelines is not included, because a pipeline appears as an example of a 'work' in subsection (b) rather than a 'building or structure' in subsection (a). In fact it is arguable that the painting of the pipeline forms part of the construction or the maintenance of it, and therefore falls within subsection (b) itself, or possibly that it is incidental to those operations and therefore falls within subsection (e).

2.2 Exclusions from the definition of 'construction operations'

Having defined the term 'construction operations' in subsection 105(1), the next subsection goes on to list exceptions. Just as there were several surprises in the classes of work that were included in the definition, many will find surprises in the exclusions as well.

'(2) The following operations are not construction operations within the meaning of this Part –
(a) drilling for, or extraction of, oil or natural gas;
(b) extraction (whether by underground or surface working) of minerals; tunnelling or boring, or construction of underground works, for this purpose;
(c) assembly, installation or demolition of plant or machinery, or erection or demolition of steelwork for the purposes of supporting or providing access to plant or machinery, on a site where the primary activity is –
  (i) nuclear processing, power generation, or water or effluent treatment, or
  (ii) the production, transmission, processing or bulk storage (other than warehousing) of chemicals, pharmaceuticals, oil, gas, steel or food or drink;
(d) manufacture or delivery to site of –
  (i) building or engineering components or equipment,
  (ii) materials, plant or machinery, or
  (iii) components for systems of heating, lighting, air-conditioning, ventilation, power supply, drainage, sanitation, water supply or fire protection, or for security or communications systems,
except under a contract which also provides for their installation;
(e) the making, installation and repair of artistic works, being sculptures, murals and other works which are wholly artistic in nature.'

Subsection 105(2)(a) excludes the drilling for, or extraction of, oil or natural gas. This is a simple provision, but is strictly limited. The construction of ancillary works or buildings is not excluded by this subsection, although other subsections are relevant.

Subsection 105(2)(b) is a little wider. Mineral extraction is excluded, and all tunnelling, boring and construction of underground works for that purpose are also excluded. Once again construction of ancillary works or buildings will not be excluded.

Subsection 105(2)(c) is substantially more complex. In order to decide whether a particular operation is excluded by this subsection, it is first necessary to establish the primary activity of the site. If that activity is nuclear processing, power generation, or water or effluent treatment, some but not all operations on that site will be excluded. Similarly operations on a site where the principal activity is the production, transmission, processing or bulk storage (other than warehousing) of chemicals, pharmaceuticals, oil, gas, steel or food or drink are potentially excluded.

It must be remembered that it is the principal activity on the site that is important. A hospital may have emergency power generation facilities, but power generation will not be the primary activity on the site. Operations falling within the definition in section 105(1) will not be excluded from the definition merely because they relate to power generation at a hospital.

Having established that the operation may be excluded because the principal activity on the site is within these rather diverse categories, it is then necessary to consider the nature of the operation itself. The assembly, installation or demolition of plant and machinery on such a site is excluded. The erection or demolition of steelwork for the purposes of supporting or providing access to plant and machinery on such a site is also excluded. The construction of buildings and civil engineering works, for example, is not excluded even where they are on a site where the principal activity is power generation.

In the Scottish case, *Homer Burgess Ltd* v. *Chirex (Annan) Ltd* (November 1999, Lord Macfadyan, Court of Session), it was held that pipework connecting various items of machinery and equipment on a site where the primary activity was the processing and production of pharmaceuticals was 'plant' and therefore excluded.

In *ABB Power Construction Ltd* v. *Norwest Holst Engineering Ltd* (August 2000, Judge Humphrey LLoyd QC) it was held that a contract for the installation of insulating cladding to boilers, ducting, silencers and other equipment in a power generation project was a contract for the installation of plant, even though the insulation was effectively only applied to the plant rather than forming part of its mechanism. Furthermore it was held that the site was one where the 'primary activity is power generation' despite the fact that power would not be generated on the site until the project was complete.

Subsection 105(2)(d) excludes the manufacture or delivery to site of components, equipment, materials, plant and machinery unless the contract also provides for their installation. This apparently simple distinction leads to some surprising results. A contract for the supply only of a pre-fabricated building is not covered by the Act. If the supplier is also required to carry out a simple erection process on site, which may account for a minimal proportion of the price, the contract is not excluded.

Finally, subsection 105(2)(e) excludes the making, installation and repair of artistic works. In order to come within this exclusion the work must be wholly artistic in nature. There is no definition of 'wholly artistic'. If the work has any function to perform or any benefit other than purely aesthetic it cannot be said to be wholly artistic.

There is power for the secretary of state to add or remove operations both from the definition section (105(1)) and from the exclusion section (105(2)) by order with Parliamentary approval. The Construction Contracts (England and Wales) Exclusion Order 1998 was made on 6 March 1998 and came into force on 1 May 1998; it is discussed in section 2.4 of this chapter.

The exclusion provisions of section 105 should be read restrictively, and the courts will be inclined to find that a particular operation is a construction operation for the purposes of the Act if it appears to be within section 105(1) unless it is clearly within the exclusion. This approach was demonstrated in *ABB Power Construction Ltd* v. *Norwest Holst Engineering Ltd* discussed above and also in *Palmers Ltd* v. *ABB Power Construction*. In the latter case, Palmers were scaffolding contractors who provided scaffolding to ABB Power as its subcontractor. The scaffolding was used by ABB in connection with its contract for the assembly and erection of a heat recovery steam generator at the Esso Fawley Co-generation Project. ABB's works clearly fell within section 105(2)(c) and were therefore not a construction operation. The works also fell within section 105(1)(b), being works forming or to form part of the land. They were only outside the definition because of the clear words of section 105(2)(c).

Scaffolding was covered by section 105(1)(e) if it was an integral part of, or preparatory to, or was for rendering complete, such operations as were previously mentioned in the subsection. The operation carried out by ABB was such an operation, even though it was then excluded by section 105(2)(c). Palmers' work was therefore within the definition, but not within a strict reading of the exclusion. Different elements of work, within the same contractual chain on the same site, can therefore be treated quite differently under the Act.

2.3 Definition of 'construction contract'

Adjudication, the process of dispute resolution created by the Act, only operates within a contractual context. Disputes arising other than under a contract are not capable of being referred to this type of adjudication unless the disputing parties agree to deal with their dispute in this way.

The Act does not affect every contract, however. It is only relevant to construction contracts as defined in section 104 of the Act. As with the definition of 'construction operations' considered above, the Act gives a broad definition and then provides a number of exceptions, either to the definition of construction contract or to the types of construction contracts to which the Act applies.

The broad definition is set out in sections 104(1) and 104(2):

'104–(1) In this Part a "construction contract" means an agreement with a person for any of the following –
(a) the carrying out of construction operations;
(b) arranging for the carrying out of construction operations by others, whether under sub-contract to him or otherwise;
(c) providing his own labour, or the labour of others, for the carrying out of construction operations.

(2) References in this Part to a construction contract include an agreement –
(a) to do architectural, design, or surveying work, or
(b) to provide advice on building, engineering, interior or exterior decoration or on the laying-out of landscape,
in relation to construction operations'

'A person' might be an individual, a body corporate or a number of individuals or bodies corporate acting in partnership or in joint venture.

If the contract provides for the carrying out of construction operations, a term itself defined in section 105, it is a construction contract and therefore potentially covered by the Act. Under many contracts the contractor is not required to carry out any construction work itself. Indeed it is a feature of some contracts that the contractor, who may even be a major construction company, expressly agrees not to carry out any such work. The contractor's role in such contracts is management. In a management contract, the contractor agrees and is required to subcontract all the work of construction involved in the project, contributing only its expertise in organising and co-ordinating the work. In a 'construction management contract' the contractor does not even operate through subcontractors. All the construction work is carried out by individual trade contractors under separate contracts between the employer and the trade contractors. The contractor's responsibility is purely to manage the work. This will generally include the division of the works between trade contract packages and selection of trade contractors. These types of contract are clearly within the ambit of subsection 104(1)(b).

The contractor does not need to be responsible for design, management or quality for the contract to be within the definition. Labour-only contracts and subcontracts are expressly contemplated by subsection 104(1)(c). Even the supply of labour by an agency is caught. When arrangements are made for such a supply, it may not be made clear to the agency whether the work to be done by the labour being supplied is a construction operation or not. For example, a structural steel contractor may request a number of steel erectors to attend a particular site to assist with the construction of a steel frame. It may not be obvious that the site's primary activity is the bulk storage of food. If that point is clear, it may not be obvious whether the purpose of the steel frame is to support plant or to support the roof of a building. If it is the former, the work will not be a construction operation because of the exclusion in section 105(2)(c), and so the contract for supply of labour will not be a construction contract. If it is the latter, the opposite will apply.

Consultancy agreements are not within the primary definition in section 104(1), but section 104(2) expands the definition by providing that many such agreements are included in references to construction contracts in Part II of the Act, which includes all the provisions of the Act relating to adjudication. The list of consultancy services to be included in such references seems to be limited but is in fact wide. Architectural work is included, but engineering work is not mentioned. On the other hand the provision of advice on engineering is included whereas the provision of advice on architecture is not mentioned. It is not easy to understand the significance of the distinction. It should be assumed that all retainers of professional design consultants in relation to construction operations as defined in section 105 are included.

The limitation to work and the provision of advice in relation to construction operations is however important. A contract for the carrying out of a structural survey of a building as part of the design of refurbishment works will be treated as a construction contract, whereas a contract for a similar survey for valuation purposes will not. A survey of railway signalling devices to establish whether or not they are safe will not be the subject of a construction contract, but a survey to establish where such devices should be installed in order to maximise safety will be.

'Surveying work ... in relation to construction operations' would seem to include the preparation of interim and final accounts by quantity surveyors on behalf of either the contractor or the employer, and so will also include the preparation of claims for loss and expense. The provision of legal services in connection with arbitration or litigation, even by a quantity surveyor, or expert witness advice, is not 'in relation to construction operations' and is therefore not included.

There is some debate about whether other contracts regularly entered into by design professionals fall within the definition. Collateral warranties are required by those with interests in buildings and other structures in order to place them in direct contractual relationships with the designers or others giving the warranty. The designer is asked to warrant that he will perform the obligations that he has undertaken under his retainer, so that the beneficiary of the warranty will have a direct right of action in the event that problems arise as a result of default on the part of the designer. Is this an agreement to carry out design work? Much will depend on the exact wording of the warranty, which varies considerably, but in most cases such a warranty will not be a construction contract for the purposes of the Act. The work of design is carried out pursuant to the original retainer. The warranty merely gives recourse to the beneficiary in the event of failure or other breach.

Designers are often required to agree novations of their original retainer from their client to another party – for example the successful tenderer for a design and build contract. A true novation merely switches the party to whom the obligation is owed, and the original contract remains effective. In such a case, if the original contract was a construction contract, it will remain so after novation. The novation agreement itself will have affected that contract, but is not itself a construction contract.

It is not immediately obvious that an agreement to act as planning supervisor for the purposes of the Construction (Design and Management) Regulations 1994 falls within the definition of a construction contract. The planning supervisor does not carry out construction operations, nor does he arrange for the carrying out of construction operations by others. He does not agree to carry out architectural, design or surveying work in the conventional sense of those words and he is not truly providing advice on building or other work in connection with construction operations. The Standard Form of Appointment of a Planning Supervisor (PS/99) published by the Royal Institute of British Architects (RIBA) does however assume that the agreement is a construction contract unless it relates to a contract with a residential occupier which would bring it within one of the exclusions considered below.

2.4 Exclusions from the definition of 'construction contracts'

Subsection 104(3) provides that a contract of employment is not within the definition of construction contracts. A contract of employment is itself defined by reference to the Employment Rights Act 1996, which at section 230(2) provides:

'"contract of employment" means a contract of service or apprenticeship, whether express or implied, and (if it is expressed) whether oral or in writing.'

Subsection 104(4) of the Housing Grants, Construction and Regeneration Act gives the Secretary of State power to add or remove types of contracts to or from the definition of construction contracts by statutory instrument. To date there have been no additions to the definition, but the Construction Contracts (England and Wales) Exclusion Order 1998 was made on 6 March 1998 and came into force on 1 May 1998, the same date as the Act itself. A similar order was made in respect of Scotland.

Under this Order, the following contracts are excluded from the definition of construction contracts:

'3(a) an agreement under section 38 (power of highway authorities to adopt by agreement) or section 278 (agreements as to execution of works) of the Highways Act 1980

3(b) an agreement under section 106 (planning obligations), 106A (modification or discharge of planning obligations) or 299A (Crown planning obligations) of the Town and Country Planning Act 1990

3(c) an agreement under section 104 of the Water Industry Act 1991 (agreements to adopt sewer, drain or sewage disposal works).'

These first three excluded contracts are agreements between land owners or developers and authorities that require certain works to be carried out before roads or sewers etc. are adopted, or as a condition of planning permission being given. They are agreements for the carrying out of construction operations, and therefore would be affected by the Act if not specifically excluded. Whilst the agreement with the authority is excluded by this order, the contract between the developer and the contractor is still subject to the Act, as of course are any subcontracts.

'3(d) an externally financed development agreement within the meaning of section 1 of the National Health Service (Private Finance) Act 1997 (powers of the NHS Trusts to enter into agreements).'

The National Health Service (Private Finance) Act 1997 was enacted because of serious concern that NHS trusts did not have power to enter into Private Finance Initiative (PFI) agreements as a means of financing their developments; it expressly empowers trusts to do so. Arguably this exclusion is unnecessary in the light of the general PFI exclusion that follows, but article 3 is concerned with agreements that are made pursuant to statutory provision and it is therefore appropriate that such agreements be included.

Article 4 of the Exclusion Order deals with PFI contracts:

'4–(1) A construction contract is excluded from the operation of Part II [of the Act] if it is a contract entered into under the private finance initiative, within the meaning given below.

(2) A contract is entered into under the private finance initiative if all the following conditions are fulfilled –
  (a) it contains a statement that it is entered into under that initiative or, as the case may be, under a project applying similar principles;
  (b) the consideration due under the contract is determined at least in part by reference to one or more of the following –
    (i) the standards attained in the performance of a service, the provision of which is the principal purpose or one of the principal purposes for which the building or structure is constructed;
    (ii) the extent, rate or intensity of use of all or any part of the building or structure in question; or
    (iii) the right to operate any facility in connection with the building or structure in question; and
  (c) one of the parties to the contract is –
    (i) a Minister of the Crown;
    (ii) a department in respect of which appropriation accounts are required to be prepared under the Exchequer and Audit Departments 1866;
    (iii) any other authority or body whose accounts are required to be examined and certified by or are open to the inspection of the Comptroller and

Auditor General by virtue of an agreement entered into before the commencement date [1 May 1998] or by virtue of any enactment;
(iv) any authority or body listed in Schedule 4 to the National Audit Act 1983 (nationalised industries and other public authorities);
(v) a body whose accounts are subject to audit by auditors appointed by the Audit Commission;
(vi) the governing body or trustees of a voluntary school within the meaning of section 31 of the Education Act 1996 (county schools and voluntary schools), or
(vii) a company wholly owned by any of the bodies described in paragraphs (i) to (v).'

The draftsman of this Order clearly found it something of a challenge to define a PFI contract. It is arguable that it was not necessary for him to have done so. A PFI contract is typically an agreement between a public body who requires the provision of services and an undertaking of some description who is prepared to provide them. It may be necessary for that undertaking to construct a building for the purpose of providing the services, but that need not be the subject of the PFI contract. Necessary or not, the exclusion is now clear, and the PFI contract will not be subject to the Act. The contracts between the provider of the services and the contractor who will construct the building from which the services will be provided is of course unaffected, and is within the Act.

Article 5 excludes agreements that primarily relate to the financing of works:

'5 – (1) A construction contract is excluded from the operation of Part II [of the Act] if it is a finance agreement, within the meaning given below.

(2) A contract is a finance agreement if it is any one of the following –
(a) any contract of insurance;
(b) any contract under which the principal obligations include the formation or dissolution of a company, unincorporated association or partnership;
(c) any contract under which the principal obligations include the creation or transfer of securities or any right or interest in securities;

(d) any contract under which the principal obligations include the lending of money;
(e) any contract under which the principal obligations include an undertaking by a person to be responsible as surety for the debt or default of another person, including a fidelity bond, advance payment bond, retention bond or performance bond.'

Once again, this exclusion seems at first reading to indicate excessive caution. How would an insurance contract find itself falling within the definition of construction contracts, even without the exclusion? It is possible that an insurance policy may require that insurance payments be expended in the reinstatement of damaged property, and perhaps it might then be argued that the policy was an agreement for the carrying out of construction operations.

A contract for construction operations that includes a parent company guarantee or other security, or indeed any of the other provisions listed in article 5(2)(b) to (e), will not be excluded from the operation of the Act merely because of the presence of that listed provision. The principal obligations must include such activity, and if the guarantee or other provision is a secondary obligation the exclusion will not affect it.

Article 6 excludes 'development agreements':

'6-(1) A construction contract is excluded from the operation of Part II if it is a development agreement, within the meaning given below.

(2) A contract is a development agreement if it includes provision for the grant or disposal of a relevant interest in the land on which take place the principal construction operations to which the contract relates.

(3) In paragraph (2) above, a relevant interest in land means –
(a) a freehold; or
(b) a leasehold for a period which is to expire no earlier than 12 months after the completion of the construction operations under the contract.'

An agreement to transfer an interest in land, as defined, is not affected by the Act, but if the development agreement provides for the parties to enter into a separate contract for construction operations that contract is not excluded. Furthermore any subcontract that involves construction operations or design work and the like in relation to such operations will be affected.

2.5 The mixed contract

'*s104(5)* Where an agreement relates to construction operations and other matters, this Part applies to it only so far as it relates to construction operations.

An agreement relates to construction operations so far as it makes provision of any kind within subsection (1) or (2).'

This limitation on the applicability of the Act may lead to serious problems of jurisdiction. The supply of materials under a contract that does not provide for their installation is not a construction operation and the Act does not apply. The same contract may however also provide for the supply of other materials that are to be installed. For example a flooring contractor may have a contract for the supply and laying of an area of carpet tiles in an office building, and also for the supply only of a quantity of similar tiles to be kept as a stock for future use as replacements. The Act applies to the first part of the contract, but not the second. Similarly a heating engineer may be asked to design a heating system and manufacture and supply components for the system for installation by others. The manufacture and supply will not be a construction operation. The design of the components will not be a construction contract. The design of the installation of the system will however relate to construction operations and will therefore be affected by the Act.

The contract may simplify matters by including an agreement for adjudication in relation to all matters under the contract, even though the Act does not require it to. In such a case a dispute about the quality of the carpet tiles being supplied for stock might be referred to adjudication as well as a dispute about the quality of the tiles being laid. Without such a provision though an adjudicator may have to split the dispute into two, and make his decision only on the part of the contract covered by the Act.

2.6 Limits of date and place

'*s104(6)* This Part applies only to construction contracts which –
    (a)   are entered into after the commencement of this Part, and

(b) relate to the carrying out of construction operations in England, Wales or Scotland.

(7) This Part applies whether or not the law of England and Wales or Scotland is otherwise the applicable law in relation to the contract.'

The date of commencement of the relevant part of the Act was 1 May 1998. Contracts entered into before this date are not affected.

Several problems have arisen in determining when a contract was entered into. A typical example came before the Technology and Construction Court in July 1999, *The Project Consultancy Group* v. *The Trustees of the Gray Trust*. One party argued that a contract had been formed on 10 July 1998 and the other said that it had been formed on 23 April 1998. Each relied on different correspondence. The defendant challenged the enforceability of the adjudicator's decision on the grounds that there was no jurisdiction. On an application for summary judgment Mr Justice Dyson had to decide whether there was an arguable defence. He decided that there was, because he was unable to decide on the basis of affidavit evidence, which was substantial, when the contract had been formed. Indeed he was not certain that there was a contract at all. If it was possible to argue that there was no contract, there was an arguable defence and summary judgment was accordingly refused.

In *Christiani & Nielsen Ltd* v. *The Lowry Centre Development Co Ltd* (Judge Thornton, June 2000) it was again argued that the contract predated May 1998. Work had started on the basis of a letter of intent dated August 1997, but the parties had executed a deed dated December 1998. It was held that the deed had superseded the letter of intent, and that therefore the relevant contract was the deed. The Act applied.

It is suggested in *Keating on Building Contracts* that it is helpful to ask the following questions when considering whether a contract has come into existence:

'(1) In the relevant period of negotiation did the parties intend to contract?
(2) At the time when they are alleged to have contracted, had they agreed with sufficient certainty upon the term which they then regarded as being required in order that a contract should come into existence?
(3) Did those terms include all the terms which, even though the

parties did not realise it, were in fact essential to be agreed if the contract was to be legally enforceable and commercially workable?

(4) Was there a sufficient indication of acceptance by the offeree of the offer as then made complying with any stipulation in the offer itself as to the manner of acceptance?'

If those questions can be answered in the affirmative, it is necessary to ask at what point it became possible to do so. The mere start on site does not necessarily indicate the date on which the parties entered into the contract. In *Trollope & Colls Ltd* v. *Atomic Power Constructions Ltd* (1961) work had started on site in June 1959 but negotiations continued and the contract did not come into existence until April 1960. It was held to have retrospective effect.

The date that appears on a contract document may indicate the date the contract was entered into, but if all the questions are satisfied at some earlier date and the dated contract does not add anything to the relationship, it may be found that the contract was entered into at the earlier date (see for example *Hatzfeld-Wildenburg* v. *Alexander* (1912)).

On the other hand the contract can be signed and dated and yet not in fact be made until later. In *The Atlas Ceiling & Partition Co Ltd* v. *Crowngate Estates (Cheltenham) Ltd* (Judge Thornton QC, March 2000), a DOM/2 subcontract had been signed on 3 April 1998. The subcontractor produced evidence that there was express agreement that the subcontract would not be entered into until various outstanding matters had been resolved. There was correspondence suggesting that this was not achieved until April 1999. There was a dispute about the final account, which the subcontractor referred to adjudication. Crowngate objected to the appointment on the basis that the contract was made on 3 April 1998, before the Act came into operation. The adjudicator found against them and decided that they should pay some £90,000 to the subcontractor. This was supported by the judge when application was made to enforce the decision.

Work frequently starts in reliance on a letter of intent. The interpretation of such a document is often difficult. Depending on the exact words of the specific letter of intent, it may operate as simply a statement of future intent with no contractual effect at all; it may be an 'if' contract entitling the recipient to payment *if* he carries out work, without any obligation on him to do so; it may be or form part of a contract for specific work other than the whole work that is to be the subject of a future contract; or it may be an acceptance of an offer and effectively the contract that it purports not to be. If no contract results from it there may or may not be an entitlement to payment on a quantum meruit. Some of these principles appear in *Turriff Construction Ltd v. Regalia Knitting Mills Ltd* (1971), *British Steel Corporation v. Cleveland Bridge Engineering Co Ltd* (1984), and *Monk Construction Ltd v. Norwich Union* (1992), although every letter of intent must be considered carefully on the basis of its own wording.

A letter of intent was the basis of an alternative argument by the defendant in *The Atlas Ceiling & Partition Co Ltd v. Crowngate Estates (Cheltenham) Ltd* (see above). The main contractor had sent a letter of intent to the subcontractor in December 1997 containing this provision:

> 'In the event that a contract is entered into between us it shall have retrospective effect to include all works carried out under this letter and you will credit us under the contract to the value of any payments made hereunder.'

Having failed to persuade the judge that the date on the contract itself was to be taken as the date it was made, Atlas argued that the contract should be deemed to have been made on the date of the letter of intent. This argument also failed. The subcontract might take effect from the date of the letter, but it was not entered into for the purposes of the Act until some 16 months later.

Section 104(6) limits the effect of Part II of the Act to England, Wales and Scotland. In fact it now also applies to Northern Ireland as a result of the Construction Contracts (Northern Ireland) Order 1997. The Act has no relevance to contracts for work done outside those limits. A contract for design works in relation to construction operations in the UK which itself involves design works elsewhere is however covered.

Section 104(7) is aimed at preventing artificial avoidance of the Act by providing that the law of the contract is that of another country. Hence an adjudicator may have jurisdiction to deal with a dispute between a contractor and a subcontractor, both from other countries, who have contracted on the basis of foreign law for construction operations in the UK. How the adjudicator's decision would be enforced is a matter for the courts in one or other of the relevant countries.

2.7 Residential occupiers

Section 106(1) provides:

'This Part does not apply –
(a) to a construction contract with a residential occupier (see below), or
(b) to any other description of construction contract excluded from the operation of this Part by order of the Secretary of State.'

Section 106(2) then defines 'residential occupier':

'A construction contract with a residential occupier means a construction contract which principally relates to operations on a dwelling that one of the parties to the contract occupies, or intends to occupy, as his residence.
In this subsection "dwelling" means a dwelling-house or a flat; and for this purpose –
"dwelling-house" does not include a building containing a flat; and
"flat" means separate and self-contained premises constructed or adapted for use for residential purposes and forming part of a building from some other part of which the premises are divided horizontally.'

A contract between a builder and a residential occupier (for example, for an extension) is not within the Act. Similarly a contract for the construction of a house between the builder and the person who intends to occupy the house is excluded. A house built for a developer or a housing association is not excluded, and contracts between the builder and subcontractors are subject to the Act.

A limited company entering into a contract for the construction of a residential property for the benefit of a director or employee is not a residential occupier and such a contract is not excluded (*Absolute Rentals Ltd* v. *Gencor Enterprises Ltd* – Judge Wilcox, July 2000).

It can be difficult to apply this to a building with a mixed use. A contract for the construction of a house with a built-in office would probably relate 'principally' to a dwelling, whereas a contract for the construction of an office block with a built-in flat would not. A contract for the conversion of a barn to an office would not be excluded but a similar conversion to a house (made with the intended occupier) would be excluded.

As with the other exclusions, there is nothing to stop the parties expressly agreeing to include an adjudication agreement in their contract. The JCT Form of Building Contract for a Home Owner/Occupier includes an adjudication agreement that provides for adjudication on a similar, but not identical, basis to adjudication under the Act.

2.8 Contracts in writing

The Act, and therefore the automatic right to refer disputes to adjudication, only applies to agreements in writing. This is provided, as are relevant definitions, by section 107 of the Act:

'107–(1) The provisions of this Part apply only where the construction contract is in writing, and any other agreement between the parties as to any matter is effective for the purposes of this Part only if in writing.

The expressions "agreement", "agree" and "agreed" shall be construed accordingly.

(2) There is an agreement in writing –
 (a) if the agreement is made in writing (whether or not it is signed by the parties),
 (b) if the agreement is made by exchange of communications in writing, or
 (c) if the agreement is evidenced in writing.

(3) Where parties agree otherwise than in writing by reference to terms which are in writing, they make an agreement in writing.

(4) An agreement is evidenced in writing if an agreement made otherwise than in writing is recorded by one of the parties, or by a third party, with the authority of the parties to the agreement.

(5) An exchange of written submissions in adjudication proceedings, or in arbitral or legal proceedings in which the existence of an agreement otherwise than in writing is alleged by one party against another party and not denied by the other party in his response constitutes as between those parties an agreement in writing to the effect alleged.

(6) References in this Part to anything being written or in writing include its being recorded by any means.'

The definition of 'in writing' is the same as that found in section 5 of the Arbitration Act 1996, save that 'adjudication proceedings' has been added to subsection (5).

A great many contracts and subcontracts in the construction industry are made without the formality of a contract document in a custom drafted or even standard form. They are made by telephone, sometimes confirmed with a purchase order and sometimes not. If there is a formal contract document, drawn up and signed, subsection 2(a) applies and it is not surprising that the Act will recognise an agreement in writing. Subsection 2(b) is also unlikely to be controversial, in that an exchange of letters or other written communications is self-evidently an agreement in writing.

'Evidenced in writing', as in subsection 2(c), is however more difficult. Sub-section 4 offers some help, in that it establishes that if the agreement is recorded in writing with the authority of the parties, the agreement is evidenced in writing. It is not clear however whether there is any other way of evidencing the contract in writing, or what is meant by 'with the authority of the parties'. If for example one party to the contract makes a note in his diary of having agreed terms (price, start date etc.), will that record be evidence in writing? There is no authority to assist, but it is submitted that the courts are likely to find that such evidence would be sufficient to bring the contract within the ambit of the Act.

Subsection (5), like its Arbitration Act cousin, is also remarkable. It suggests that the agreement giving rise to the dispute may have been oral, with no writing at all, until adjudication proceedings are started, and then metamorphose into an agreement during the course of those proceedings because the respondent does not deny the existence of the (until then) oral contract. This raises the question as to how the proceedings can have been started in reliance on an implied adjudication provision that cannot have been in existence until part of the way through the proceedings themselves.

This conundrum was considered by the Technology and Construction Court in February 2000 in *Grovedeck Ltd* v. *Capital Demolition Ltd*. Grovedeck were subcontractors to Capital Demolition for demolition works on two sites. The contracts for each were made orally. Disputes arose and an adjudicator was appointed. Capital Demolition asked the adjudicator to consider whether or not he had jurisdiction as a preliminary point. They argued that there was no material for finding a contract in writing under any of subsections 107(2)(a)–(c). Grovedeck were unable to find any contemporaneous documents evidencing the contract, but relied initially on correspondence written after the disputes arose.

The adjudicator decided that there were 'valid construction contracts' and pressed on with the adjudication, requiring Capital Demolition to serve a response. In that document, Capital Demolition denied that the contracts were in writing or evidenced in writing, and having reserved their position on that, went on to comment on the claim itself. They did not deny the existence of the contracts, but argued they were not in writing.

The adjudicator decided in favour of Grovedeck. In his decision, which was accompanied by reasons, he found that Grovedeck had alleged an oral contract, and Capital Demolition had not denied it. Therefore, he said, section 107(5) applied and there was indeed an agreement in writing.

Grovedeck sought to enforce the decision. In the enforcement proceedings Grovedeck abandoned the post dispute correspondence as evidence of the contract in writing and adopted the argument based on subsection (5). Judge Bowsher was unable to accept that:

'The contracts were not subject to any terms about adjudication when the adjudicator was appointed and so, at the date of his appointment, he had no jurisdiction. Did something happen later to change the nature of the contracts between the parties and give jurisdiction to the adjudicator so as to bestow validity on what was proceeding as an invalid adjudication? The claimants say, Yes. The claimant's submissions involve this unstated proposition that even though in every communication after his unlawful appointment the defendants challenged and denied the jurisdiction of the adjudicator, those same communications themselves changed the nature of the parties' contracts and gave him jurisdiction. Freedom of contract has fallen, but I cannot believe that it has fallen that far.'

Having considered the passages from *Hansard* dealing with the introduction of the reference to adjudication proceedings in subsection (5), Judge Bowsher concluded that the subsection really only referred to '*other preceding* adjudication proceedings' (the judge's emphasis).

The contract does not need to be complete in all its terms before it will be treated as a contract in writing for the purposes of the Act. In *R.G. Carter Ltd* v. *Edmund Nuttall Ltd* (June 2000), Judge Thornton was asked to prevent the prosecution of an adjudication by a subcontractor against a main contractor. The subcontractor's tender had been accepted by the main contractor, subject to a set of amendments to the standard form DOM/1. Those amendments included a mandatory provision for submission of any dispute to mediation before an adjudication could be commenced. There was no evidence that the subcontractor had accepted those amendments, or any of them. The judge held that there was a contract in writing sufficient to satisfy the requirements of the Act, although the parties might not be agreed as to all of its terms. In any event the requirement to submit disputes to mediation was contrary to the statutory requirement that the parties be able to refer disputes to adjudication 'at any time'.

CHAPTER THREE
THE STATUTORY RIGHT TO REFER DISPUTES TO ADJUDICATION

'108(1) A party to a construction contract has the right to refer a dispute arising under the contract for adjudication under a procedure complying with this section.

For this purpose "dispute" includes any difference.'

The Act gives any party to a construction contract the right to refer a dispute to adjudication. This right is not itself a term to be implied into the contract, although the mechanism of the adjudication process will depend on implied terms to make it effective. The right is a statutory right which exists whatever the contract says about it. Section 108(1) of the Act expresses this in extremely simple and direct terms.

3.1 Definition of dispute

The Act anticipates that there may be some difficulty in establishing whether a dispute has arisen. If it is possible to argue that a matter has been referred for adjudication that was not in fact a dispute there would be serious doubt about whether the decision when made is enforceable. The losing party would object that the decision is not a valid adjudicator's decision at all, and therefore not something on which the winning party can rely. It is therefore stated that the word 'dispute' includes any difference. Questions then arise as to the meaning of the word 'difference'.

These questions have been debated in several cases involving arbitration and the applicability of agreements to arbitrate. In summary, it seems that the courts will be slow to conclude that a dispute or difference does not exist if the parties have found it necessary to embark on litigation or arbitration.

In *Cruden Construction Ltd* v. *Commission for the New Towns* (1995) Judge Gilliland reviewed older cases and summarised them by saying that if a claim was raised and ignored or met with prevarication a dispute could be said to exist. Nevertheless he declined to find any general rule as to the meaning of the word.

The point has taken on particular significance since the Arbitration Act 1996 came into force. If a contract contains an agreement to refer disputes to arbitration, and despite that agreement one party chooses to issue court proceedings in relation to a matter covered by the arbitration agreement, the other party can apply to the court to stay those court proceedings so that the matter can be dealt with by arbitration. Under section 9 of the Arbitration Act 1996 the court has no discretion in the matter. A stay will be granted unless the court is satisfied that the arbitration agreement is null and void, inoperative or incapable of being performed.

A party seeking to avoid arbitration and continue with court proceedings is therefore obliged to argue that the arbitration agreement does not in fact apply to the matter in question. A potential basis for this argument is that there is no dispute. That was the approach taken by the plaintiff in *Halki Shipping Corporation* v. *Sopex Oils Ltd* (1997). It claimed demurrage payments under a charterparty. The defendant did not admit liability, but neither did it put forward any reason for not paying. It simply relied on an arbitration agreement in the charterparty, and sought a stay of the court proceedings.

The Admiralty Judge followed the approach of Mr Justice Saville in *Hayter* v. *Nelson and Home Insurance Co* (1990). He had examined the proposition that an indisputable claim could not be referred to arbitration and found it to be unsustainable. The logical conclusion would be that if an arbitrator found that there was no basis for defending a claim he would have no jurisdiction to make an award. This was absurd. Indisputable matters could form the basis of disputes:

> 'In my judgment in this context neither the word "disputes" nor the word "differences" is confined to cases where it cannot then and there be determined whether one party or the other is in the right. Two men have an argument over who won the University Boat Race in a particular year. In ordinary language they have a dispute over whether it was Oxford or Cambridge. The fact that it can be easily and immediately demonstrated beyond any doubt that one is right and the other is wrong does not and cannot mean that that dispute does not in fact exist. Because one man can be said to be indisputably right and the other indisputably wrong does not, in my view, entail that there was therefore never any dispute between them.'

If therefore there is no statement in the contract as to what is meant by dispute or difference it will be difficult for a reluctant party to argue that a matter should not be referred to adjudication. Complications may arise if there is an attempt in the contract to supplement the Act by defining the circumstances in which a dispute exists. Such an attempt may be a deliberate move to frustrate the intention of the statute to provide a quick and relatively inexpensive means of dispute resolution. Alternatively it may be based on a more honourable motive, encouraging parties to use a different resolution process that is made available to them. An example of the latter is found in the ICE forms of contracts and subcontracts.

The ICE 7th Edition Main Contract includes a comprehensive procedure for avoiding and settling disputes. Clause 66(2) provides that if the contractor or employer is dissatisfied with any of a range of matters the 'matter of dissatisfaction shall be referred to the Engineer who shall notify his written decision' to the parties within one month.

Clause 66(3) then provides:

'The Employer and the Contractor agree that no matter shall constitute nor be said to give rise to a dispute unless and until in respect of that matter
(a) the time for the giving of a decision by the Engineer on a matter of dissatisfaction under Clause 66(2) has expired or the decision given is unacceptable or has not been implemented and in consequence the Employer or the Contractor has served on the other and on the Engineer a notice in writing (hereinafter called the Notice of Dispute)...'

The only exception to this is the case of an adjudicator's decision with which a party has not complied.

Clause 66(3) concludes thus:

'For the purposes of all matters arising under or in connection with the Contract or the carrying out of the Works the word "dispute" shall be construed accordingly and shall include any difference.'

The Civil Engineering Contractors Association (CECA) form of subcontract, designed to be used with the ICE 6th Edition Main Contract, includes a similar limitation in clause 18(2)(b):

'...The Contractor and the Sub-Contractor that no such submission [that additional payment is due] shall constitute nor be said to give rise to a dispute under the Sub-Contract unless and until the Contractor has had the time and opportunity to refer the matter of dissatisfaction to the Engineer under the Main Contract and either the Engineer has given his decision or the time for giving a decision by the Engineer has expired.'

The equivalent wording of the New Engineering and Construction Contract as amended by Option Y(UK)2, purportedly to comply with the Act, is even more clear:

'90.4 The Parties agree that no matter shall be a dispute unless a notice of dissatisfaction has been given and the matter has not been resolved within four weeks. The word dispute (which includes a difference) has that meaning.'

Such limitation may well be effective in requiring parties to employ particular procedures before embarking on arbitration proceedings. Arbitration is after all a contractual process and is only available to the parties because they have agreed that it should be available. In an exclusively contractual arrangement the parties are at liberty to agree that words have particular meanings within the context of their contract.

The right of parties to refer disputes to adjudication is not however a contractual right. It is a statutory right, and it exists regardless of the expressed terms of the contract. It is not possible to restrict the effect of the Act by defining the word 'dispute' within the contract, even though it is possible to restrict the availability of arbitration procedures by that means. If a difference, as normally understood, arises between two parties to an ICE contract it is possible for one party to refer it to adjudication even if the other procedures have not been followed.

The adjudication provisions of the New Engineering Contract (otherwise known as the Engineering and Construction Contract), set out in part above, were considered in *John Mowlem plc* v. *Hydra-Tight Ltd* (August 2000). Judge Toulmin considered that the clause did not provide the parties to the contract with the right to refer disputes to adjudication at any time. As a result the contractual adjudication procedures were non-compliant, and the Scheme adjudication procedures applied in substitution.

In *R.G. Carter Ltd* v. *Edmund Nuttall Ltd* (Judge Thornton, June 2000) a main contractor applied to court to prevent a subcontractor from proceeding with an adjudication. It contended that the subcontract had included an agreement to take disputes to a mediation procedure before adjudication, and to appoint a named adjudicator rather than a person selected by an adjudicator nominating body. The application failed through lack of certainty that these provisions had been incorporated into the contract, but the judge also held that the mandatory mediation procedure sought to fetter the unqualified entitlement to adjudication provided by the Act and was therefore unenforceable.

It is not every dispute or difference between two parties to a construction contract that can be referred to adjudication. Section 108(1) refers only to 'a dispute arising under the contract'. If the dispute arises other than under the contract it is not subject to the statutory right to refer.

There is little case law to assist in defining 'under the contract' specifically in relation to adjudication, but there have been disputes in the past about the ambit of arbitration clauses. A variety of phrases have been used, all of which seem to have subtly different scopes. 'Disputes arising under a contract' has been held not to be wide enough to cover disputes which do not involve obligations created by or incorporated in the contract (*Fillite* (Runcorn) v. *Aqua-Lift* (1989)), whereas the inclusion of the words 'or in connection therewith' has been held to enable the arbitrator to deal with claims relating to mistake, leading to a claim for rectification (*Ashville Investments* v. *Elmer Contractors Ltd* (1987)).

As the Act only requires the contract to provide for disputes arising under the contract to be referred to adjudication, there is no right to refer disputes in these wider categories, unless of course the contract expressly so provides.

A claim that the contract has been repudiated does arise 'under the contract'. This point arose in *Northern Developments (Cumbria) Ltd* v. *J. & J. Nichol* (Judge Bowsher, January 2000). It was suggested that because there had been an 'accepted' repudiation, there was no contract capable of being considered by the adjudicator. Judge Bowsher applied the law developed in relation to arbitration agreements, and in particular the House of Lords decision in *Heyman* v. *Darwins* (1942). He quoted the words of Lord Russell of Killowen:

> 'Repudiation, then, in the sense of a refusal by one of the parties to a contract to perform his obligations thereunder does not of itself abrogate the contract. The contract is not rescinded. It obviously cannot be rescinded by the action of one of the parties alone. But even if the so-called repudiation is acquiesced in or accepted by the other party, that does not end the contract. The

wronged party has still his right of action for damages under the contract which has been broken, and the contract provides the measure of those damages.'

The House of Lords had concluded that repudiation, or indeed frustration, brings the performance of the contract to an end, but not the contract itself. Accordingly the arbitration clause would survive. The same was true of adjudication agreements.

Judge Wilcox applied the same logic in *A. & D. Maintenance and Construction Ltd v. Pagehurst Construction Ltd* (June 1999). The unsuccessful contractor had sought to argue that as the subcontract had been determined there was no subcontract in existence, and therefore a dispute could not arise under it. Judge Wilcox had also referred to *Heyman* v. *Darwins* in rejecting the argument.

A dispute that arises under a compromise agreement does not arise under a construction contract, even if the original dispute, which had been resolved by the compromise agreement, was itself a construction contract. In *Lathom Construction Ltd* v. *Cross and Cross* (Judge Mackay, October 1999) a contract dispute had been referred to adjudication and then settled. A dispute then arose about the terms of compromise, and the contractor applied to the RICS for the appointment of an adjudicator in respect of that dispute. The same adjudicator was appointed and found in favour of the contractor. The employer successfully resisted enforcement on the basis that there was a genuine dispute about the jurisdiction of the adjudicator.

A similar point arose in *Shepherd Construction Ltd* v. *Mecright Ltd* (Judge LLoyd, July 2000). Mecright started an adjudication to recover the value of its works. Shepherd applied to court for a declaration that there was in fact no dispute to refer to adjudication because there had already been a settlement agreement. Mecright alleged that the agreement had been procured as a result of economic duress, but it was held that before Mecright could show that there was a dispute under the construction contract it would have to have the settlement agreement set aside by the court. At the time of referral to adjudication of the dispute as to value this had not been done, and therefore there was no dispute capable of being referred. A dispute as to whether or not a settlement was binding was not a dispute under a construction contract.

Several standard form contracts go further than the strict requirements of the Act by enabling a wider class of disputes to be referred to adjudication than just disputes 'arising under the contract'. For example, the relevant clause of the Government contract GC/Wks/1 reads:

'59(1) The Employer or the Contractor may at any time notify the other of intention to refer a dispute difference or question arising under, out of, or relating to, the Contract to adjudication.'

Part II of the Act only ever deals with the word 'contract' in the singular. Nevertheless the right to refer a dispute to adjudication may apply where the dispute has arisen under more than one contract. This is the effect of sections 5 and 6 of the Interpretation Act 1978, which provides that unless a contrary intention appears, words in the singular include words in the plural. A contrary intention does not appear in the Housing Grants, Construction and Regeneration Act. This does not mean that there is an automatic right to refer disputes under several contracts to adjudication before the same adjudicator. The rules of the relevant adjudication provisions in the contracts may govern the point, as they do in the Scheme (paragraph 8). Judge Bowsher dealt with this point in *Grovedeck Ltd* v. *Capital Demolition Ltd* (February 2000):

'I see no reason why a construction contract in writing which sufficiently complied with section 108 of the Act as to avoid the application of the Scheme should not provide for the referral of more than one dispute or more than one contract without the consent of the other party. Parties might be unwise to agree to such a term, but I do not see why they should not do so. Section 108(2)(a) of the Act requires that a construction contract shall "enable a party to give notice at any time of his intention to refer a dispute to adjudication" but I do not read that as showing any intention that the singular does not include the plural.'

As the Scheme applied to the Grovedeck contract, only one dispute under one contract could be referred without the consent of the other party.

3.2 Required contractual provisions

Having established that there is a dispute to which the Act applies and that there is therefore a right to refer it to adjudication, the question to be considered is how the adjudication is to be conducted. The Act itself does not provide an answer. The Act merely sets out a set of minimum criteria by which the contractual machinery is to be judged. If the contract provides a set of procedural rules that satisfy those criteria, those rules will not suffer interference from the legislation. If however they fail any of the criteria, the adjudication provisions of the Scheme will replace the contractual rules for construction contracts.

The criteria are set out in sections 108(2) to 108(4) of the Act:

> '108–(2) The contract shall –
> (a) enable a party to give notice at any time of his intention to refer a dispute to adjudication;
> (b) provide a timetable with the object of securing the appointment of the adjudicator and referral of the dispute to him within 7 days of such notice;
> (c) require the adjudicator to reach a decision within 28 days of referral or such longer period as is agreed by the parties after the dispute has been referred;
> (d) allow the adjudicator to extend the period of 28 days by up to 14 days, with the consent of the party by whom the dispute was referred;
> (e) impose a duty on the adjudicator to act impartially; and
> (f) enable the adjudicator to take the initiative in ascertaining the facts and the law.
>
> (3) The contract shall provide that the decision of the adjudicator is binding until the dispute is finally determined by legal proceedings, by arbitration (if the contract provides for arbitration or the parties otherwise agree to arbitration) or by agreement.
>
> The parties may agree to accept the decision of the adjudicator as finally determining the dispute.
>
> (4) The contract shall also provide that the adjudicator is not liable for anything done or omitted in the discharge or purported discharge of his functions as adjudicator unless the act or omission is in bad faith, and that any employee or agent of the adjudicator is similarly protected from liability.'

3.2.1 'at any time'

The first criterion is that the contract must enable a party to give notice at any time of his intention to refer a dispute to adjudication. We have already seen that the ICE family of contracts and subcontracts seek to fetter the ability of a party to refer a dispute by providing an artificial definition of 'dispute'. These contracts do not enable a party to give notice of intention to refer at any time, and accordingly the contracts fail to meet the criteria. This has serious consequences for the rules for adjudication procedure set out in those contracts. The Scheme for Construction Contracts applies automatically and the adjudicator will have to consider the extent to which the ICE rules apply to supplement the Scheme. Where the rules conflict, the Scheme will have priority.

The required ability to refer disputes at any time means that if contractual rules are to comply, they must allow an adjudication to be started long after the completion of the works that formed the subject of the contract. It may be that the issue of a final certificate will have conclusively established the rights of the parties, but the parties will still be at liberty to commence adjudication to ask the adjudicator what those rights are. The adjudicator will not have any better right to open up a final certificate than the court or an arbitrator.

The right to refer a dispute to adjudication survives the determination of the contract. This has long been established in the case of arbitration, but the contrary was argued, unsuccessfully, in *A. & D. Maintenance and Construction Ltd* v. *Pagehurst Construction Services Ltd* (Judge Wilcox, June 1999):

> 'Had it been the intention of Parliament to limit the time wherein the party could give notice of his intention to refer a matter to adjudication, in the exercise of his right under s108(1), it could have imposed a clear time limit. Precise limits as to the appointment of adjudicators and the timetabling of the process of adjudication are clearly set out in the Scheme. By contrast there is no such limitation under the Act or the Scheme as to when a notice of intention to refer a matter to adjudication may be made. By analogy with arbitration provisions, there is clear authority to the proposition that those terms governing reference to arbitration survive the determination of the contract.'

It is a common feature of many contractors' standard forms of subcontract that the subcontractor is not able to start arbitration proceedings until after practical completion of the main contract works. An attempt to limit the subcontractor's rights to start adjudication proceedings in the same way will fail. Such a provision will mean that the contractor's rules for adjudication proceedings do not satisfy the criteria required and they will be replaced by the Scheme for Construction Contracts.

There is an effective limitation on the ability of a party to commence adjudication proceedings 'at any time' in section 11(3)(d) of the Insolvency Act 1986, which provides:

'During the period for which an Administration Order is in force no other legal proceedings and no execution or other legal process may be commenced or continued and no distress may be levied against the company or its property except with the consent of the administrator or the leave of the court and subject, (where the court gives leave) to such terms as aforesaid.'

In *A. Straume (UK) Ltd* v. *Bradlor Developments Ltd* (1999) the plaintiff wished to bring adjudication proceedings against a company in administration. The latter company had already started an adjudication in respect of certified sums. Straume wished to pursue set-off claims. The application for leave was heard in the Chancery Division. It was held that adjudication was 'other legal proceedings'. Judge Behrens said:

'I have come to the clear conclusion that the adjudication procedure under section 108 of the Act and/or clause 41 [of JCT80] is quasi legal proceedings such as arbitration... It seems to me that it is, in effect, a form of arbitration, albeit the arbitrator has a discretion as to the procedure that he uses, albeit that the full rules of natural justice do not apply.'

Accordingly leave was required before adjudication proceedings could be started. In the circumstances of that case, leave was refused.

The position would be the same in the case of a company being wound up by the court. Section 130(2) of the Insolvency Act 1986 provides:

'When a winding-up order has been made or a provisional liquidator has been appointed, no action or proceeding shall be proceeded with or commenced against the company or its property, except by leave of the court and subject to such terms as the court may impose.'

Section 5 of the Limitation Act 1980 provides that in the case of a simple contract, not signed as a deed, the time limit for bringing legal proceedings is six years from the date on which the cause of action arose. If the contract is signed as a deed (or 'under seal'), the limit is 12 years (section 8 of the Limitation Act). That time runs from the date of the breach which gives rise to the cause of the action. The date of discovery of the breach is not relevant.

In a building contract, when work may be carried out over a

period of many months, it can be difficult to establish the date of the breach. It is suggested that the time runs from the last date on which the contract could have been performed, even if the defective work was in fact carried out earlier. Typically therefore the period will be calculated from the date of practical completion.

The Limitation Act 1980 however does not directly affect the commencement of adjudication proceedings. The time limits described above apply to actions in court, not to other types of action. The same limits were specifically extended to arbitration proceedings by section 34(1) of the Limitation Act, and subsequently by section 13 of the Arbitration Act 1996. If section 5 of the Limitation Act required specific statutory provision before it applied to arbitration proceedings, a similar statutory provision would be required to apply section 5 to adjudication proceedings. There is therefore no reason why an adjudication cannot be started more than six years after the accrual of the cause of action. That of course is unlikely to be relevant until after 1 May 2004, the sixth anniversary of the birth of statutory adjudication.

This does not mean that the Housing Grants, Construction and Regeneration Act has effectively abolished limitation of actions in construction contracts arising after 1 May 1998. The adjudicator's decision still has to be enforced. The legal action or arbitration proceedings taken to enforce the decision will be subject to the Limitation Act. The question then arises as to when the cause of action arose. A claimant facing a potential limitation defence will argue that the cause of action is the adjudicator's decision, not the date on which the claim itself arose. The adjudicator may have stated when the sum of money became due, and bearing in mind that the adjudicator's decision is binding until finally determined by legal proceedings, arbitration or agreement, this statement will have some force. It is possible that the adjudicator will decide that the money became due many years after the currency of the contract, and even if he is wrong his decision will be enforceable (following the logic of *Bouygues UK Ltd* v. *Dahl-Jensen UK Ltd*, November 1999). If the courts are faced with this problem the approach will probably be to treat the cause of action as being the original claim that had been the subject of the adjudication, not the decision itself, or alternatively to proceed straight to a trial of the issue of limitation at the hearing of the application to enforce. A more certain outcome could be achieved by legislation, introducing a provision that adjudication proceedings are subject to the operation of the Limitation Act in the same way as arbitration.

3.2.2 'provide a timetable'

The second criterion by which a contract's adjudication machinery will be judged is whether it provides a timetable with the object of securing that appointment of the adjudicator and referral of the dispute to him within seven days of the notice of intention to refer. The machinery must be designed to enable this to be achieved, but failure in practice to meet the seven day target does not invalidate the procedure.

If the adjudicator is named in the contract it is easy to provide the requisite timetable. After all, the claimant only needs to send the notice of referral to the adjudicator, and that can be done immediately upon giving notice of intention to refer.

If there is no adjudicator named in the contract one will have to be selected and appointed. This is likely to take a substantial part of the seven day period. Indeed experience of dealing with bodies charged with the task of appointing or nominating arbitrators gave rise to concern that the timetable would seldom be met.

The standard forms of contract incorporate a timetable but do not suggest how the timetable can be achieved. The JCT Standard Form of Contract 98 contains this provision:

> '41A.2.2 where either Party has given notice of his intention to refer a dispute to adjudication then
> - any agreement by the Parties on the appointment of an Adjudicator must be reached with the object of securing the appointment of, and the referral of the dispute or difference to, the Adjudicator within 7 days of the date of the notice of intention to refer (*see clause 41A.4.1*);
> - any application to the nominator must be made with the object of securing appointment of, and the referral of the dispute or difference to, the Adjudicator within 7 days of the notice of intention to refer.'

The ICE 7th Edition is equally unhelpful:

> '66(6)(b) Unless the adjudicator has already been appointed he is to be appointed by a timetable with the object of securing his appointment and referral of the dispute to him within 7 days of such notice.'

Whilst these standard provisions do not provide guidance as to how the timetable is to be achieved, they do satisfy the requirement of providing a timetable. Compliance with that timetable is then a matter for the parties and for the adjudicator nominating bodies. In practice the adjudicator nominating bodies are acting very promptly upon receipt of a request to nominate. The appointment is typically made within 48 hours.

Failure to provide a timetable, at least in a rudimentary form, will mean that the Scheme for Construction Contracts will apply and other contractual provisions regarding adjudication will only be effective insofar as they do not conflict with the Scheme.

3.2.3 'a decision within 28 days'

The contract must require the adjudicator to reach a decision within 28 days of the date of referral, subject to limited rights of extension of time. This time runs not from the date of appointment of the adjudicator, but from the date of referral of the matter to him, which (if it is achieved within the timetable provided by a compliant contract) may be as much as seven days after the notice of intention to refer. Bank holidays are not included in the computation of time by virtue of section 116, which states:

> '116–(1) For the purposes of this Part periods of time shall be reckoned as follows.
>
> (2) Where an act is required to be done within a specified period after or from a specified date, the period begins immediately after that date.
>
> (3) Where the period would include Christmas Day, Good Friday or a day which under the Banking and Financial Dealings Act 1971 is a bank holiday in England and Wales or, as the case may be, in Scotland, that day shall be excluded.'

It should be noted that whilst bank holidays are excluded, weekends are not.

If the contract does not contain this requirement, it will not comply with the Act and the Scheme for Construction Contracts will apply.

3.2.4 Extension of 'up to 14 days'

There are two possibilities for extension of time that are to be incorporated into the contract. The first extension is to be entirely within the discretion of the party referring the dispute. The adjudicator does not have power to grant himself such an extension without the consent of that party. The contract must provide that any other extension can only be given with the agreement of all the parties, and that such agreement must be made after the dispute has been referred. This prevents the inclusion in the contract of a term automatically extending the time for the adjudicator to reach his decision. Any contract that does include such an automatic extension will fail to comply with the Act's requirements and the Scheme for Construction Contracts will therefore automatically become the relevant machinery for adjudication.

3.2.5 'impose a duty on the adjudicator to act impartially'

The contract must impose a duty on the adjudicator to act impartially. This duty is equivalent to the duty imposed on arbitrators under section 33 of the Arbitration Act 1996. The contract between the parties must contain this provision, but of course that contract will not bind the adjudicator (save by implication) unless it is repeated in his appointment. The object of this criterion is to ensure that one party cannot effectively control the adjudication by providing for the appointment of a biased adjudicator. This does not mean that the adjudicator cannot be employed by or be otherwise connected with a party. Such connection may make it difficult for the adjudicator to demonstrate his impartiality, but the appointment of an adjudicator with such a connection would not be a failure to comply with the requirements of the Act, providing there is an express duty on the adjudicator to act impartially.

3.2.6 'enable the adjudicator to take the initiative'

The contract must expressly state that the adjudicator is able to take the initiative in ascertaining the facts and the law. Adjudicators in practice find that this power is essential if a decision is to be reached in just 28 days. The practical implications of this are examined later in this book. For the purposes of this chapter it is sufficient to note that unless this is provided, the Scheme for Construction Contracts will apply in the place of other provisions of the contract. Indeed if the contract does give the wide power required, and then seeks to limit the power by restricting the rights of the adjudicator to exercise his initiative in particular ways, there is a danger that the whole machinery will fail and be replaced by the Scheme.

The standard forms of contract go further than the strict requirement of the Act, emphasising the extent of the discretion that the adjudicator may exercise while taking the initiative.

3.2.7 'binding until the dispute is finally determined'

It must be a term of the contract that the decision of the adjudicator is binding until the dispute is finally determined by legal proceedings, arbitration or agreement. Such a term gives the process of adjudication real significance, as without it the losing party would be able to continue disputing matters indefinitely. As with the other requirements, a failure to include this term will invalidate the contractual machinery for adjudication and the Scheme will apply.

Section 108(3) also provides an option for the parties to accept the decision of the adjudicator as finally determining the dispute. Agreement to this effect can be made in advance either as a term of the contract, or specifically in connection with the issue in dispute when the dispute arises. The parties are unlikely to agree that a decision is final after the result, when one or other party is clearly the loser, but it is theoretically possible.

It is possible that an agreement in advance that an adjudicator's decision will be final effectively converts the adjudication process into an arbitration, which is then subject to the Arbitration Act 1996. The Arbitration Act does not contain a definition of the word 'arbitration', but the team of authors of the 'General Principles' section of *The Handbook of Arbitration Practice* (3rd edn) – Ronald Bernstein, Derek Wood, John Tackaberry and Arthur Marriott – define arbitration thus:

> 'In English law, arbitration is a mechanism for the resolution of disputes which takes place, usually in private, pursuant to an agreement between two or more parties, under which the parties agree to be bound by the decision to be given by the arbitrator according to law or, if so agreed, other considerations, after a fair hearing, such decision being enforceable at law.'

It is difficult to see how an adjudication, which the parties have agreed will produce a permanently binding result, fails to fall within this description. If such an adjudication is in fact an arbitration, the Arbitration Act will apply to it. This would have serious implications for procedure and enforcement. The Housing Grants, Construction and Regeneration Act would cease to apply to such proceedings. To make matters more confusing, however, the contract would no longer have a right to refer disputes to adjudication, and the Act would therefore impose a further adjudication process through the Scheme.

3.2.8 Liability of the adjudicator

Section 108(4) of the Act requires that the contract shall provide that the adjudicator is not liable for anything done or omitted in the discharge or purported discharge of his functions as adjudicator unless the act or omission is in bad faith, and that any employee or agent of the adjudicator is similarly protected. This provision is not a statutory immunity, such as that enjoyed by arbitrators under section 29 of the Arbitration Act 1996. Section 108(4) does no more than require parties to a construction contract to provide the adjudicator with an immunity. If they fail to do so, the Scheme will apply, but once again all that is achieved is a term of a contract between the contracting parties that the adjudicator will be immune.

The adjudicator's relationship with the parties is also contractual, but of course that relationship is governed by the contract between the parties and the adjudicator, not directly by the contract between the parties themselves. The adjudicator is not a party to that contract. If a claim is made against the adjudicator for a breach of his contract with the parties the adjudicator may be able to rely on a specific provision of his own appointment, such as that contained in the Construction Industry Council Model Adjudication Agreement or the JCT Standard Form of Adjudicator's Agreement. Alternatively he may be able to rely on the construction contract agreement as being for his benefit, under the Contracts (Rights of Third Parties) Act.

This will not necessarily be the case, and many construction contracts expressly exclude all third party benefits. It is likely that if any claim is made against an adjudicator in circumstances that do not involve bad faith, an argument will be raised that the adjudicator should be immune as a matter of public policy, and it will be said that effect should be given to the intention of the legislature, but until such an approach has been approved judicially there is some uncertainty. This is another matter that could be clarified by amending legislation.

It is difficult to know exactly what is meant by 'bad faith'. It is certainly something more serious than negligence. It has been described as 'malice or knowledge of absence of power to make the decision in question' (*Melton Medes Ltd and Another* v. *Securities and Investment Board* (1995)). There are, however, further potential limitations to the adjudicator's indemnity considered in Chapter 7, in which the indemnities set out in the published adjudication rules are discussed.

The immunity of the adjudicator being a contractual matter, third parties who are not in contract with him may have a potential cause of action in tort. It is possible to imagine a situation in which the adjudicator effectively makes a decision involving design issues with disastrous consequences. In such cases however it will be necessary to demonstrate that the adjudicator owed a duty of care to the person suffering injury. Once again, any such claim will probably be met with a defence based on public policy.

3.3 The incorporation of institutional rules and other terms

If a construction contract complies with section 108(1)–(4), it can also contain other specific provisions regarding adjudication. If those provisions do not conflict with the required terms, they will be contractually valid.

There are many institutional sets of rules that go rather further than the required matters. The detail of the principal sets will be considered later in chapters dealing with the practical process of adjudication. It is however appropriate to consider here the effect of provisions introduced into construction contracts with the apparent intent of rendering the adjudication process ineffective.

As discussed above, some construction contracts attempt to limit the time at which adjudication proceedings can be started or provide a timetable that is not compliant with section 108. Such terms, clearly not in accordance with section 108(1), cause the Scheme for Construction Contracts to be brought into the contract, and are therefore ineffective.

Terms of the contract that do not contravene the requirements of section 108 can however be very limiting. It is not contrary to the Act to provide, for example, that the referring party will always meet the costs of adjudication. Such a clause was upheld by Judge Mackay in *Bridgeway Construction Ltd* v. *Tolent Construction Ltd* (April 2000). Some contractor's standard forms require that the subcontractor shall be obliged to indemnify the contractor against any losses or costs incurred in complying with an adjudicator's decision that is subsequently reversed in arbitration or litigation. Some provide that any money that an adjudicator decides is due should be paid into a stakeholder's account until practical completion of the contract or some other time. Whether such a clause is enforceable is a matter of some debate.

3.4 The adjudication provisions of the Scheme for Construction Contracts

It is an automatic consequence of failure to comply with the requirements of section 108(1)–(4) of the Act that the adjudication provisions of the Scheme for Construction Contracts apply. The Scheme is not just used to supplement the contractual provisions, and fill in the gaps. All the adjudication provisions of the Scheme apply, including the provisions that deal with wider procedural issues than those required by section 108. They may overrule provisions set out in the contract, some of which did comply with the Act.

The detail of the provisions will be considered in subsequent chapters. Effect is given to them by section 114(4) of the Act, which provides that:

> 'Where any provisions of the Scheme for Construction Contracts apply by virtue of this Part [of the Act] in default of contractual provision by the parties, they have effect as implied terms of the contract concerned.'

The Scheme has not been amended in any way since the Act came into force on 1 May 1998, but as it is a statutory instrument rather than part of the Act itself it can be amended by regulation by the appropriate minister with approval of Parliament.

CHAPTER FOUR
STARTING ADJUDICATION

4.1 Timing

The Act requires the contract to enable a party to give notice *at any time* of his intention to refer a dispute to adjudication (section 108(2)(a)). If a construction contract does not satisfy the requirements of the Act, the Scheme for Construction Contracts will apply. The Scheme does not attempt to limit in any way the time for referring a matter for adjudication. Paragraph 1(1) simply states:

> 'Any party to a construction contract (the "referring party") may give written notice (the "notice of adjudication") of his intention to refer any dispute arising under the contract, to adjudication.'

As there is no limitation, the notice can be given at any time. There is no pre-condition. It is not necessary to adopt any other procedure before referring the matter. Clearly there must be a dispute, but section 108(1) states that the word 'dispute' includes 'any difference'. A simple dispute as to the amount that should be paid as an instalment, the value of a variation, the legitimacy of an instruction or the quality of an item of work can be referred without extensive prior correspondence explaining each side's position.

The notice can be given during the currency of the contract, while works are being carried out, or after practical completion. It is conceivable that notice might be given even before work has started. If the contract states that a decision or certificate is final and conclusive, the adjudicator will not have power to open it up or revise it, and there will therefore be little point in commencing adjudication proceedings in respect of it, but if a party wishes to start such pointless proceedings, there is nothing to stop him from doing so.

The Limitation Act 1980 does not apply directly to adjudication proceedings, although it may have indirect application by its effect on enforcement proceedings in either litigation or arbitration. This is considered in Chapter 3.

In practice the timing of the notice of adjudication is really very sensitive. It is after all the first step in a very fast moving process. The appointment of the adjudicator will follow swiftly. The detailed referral notice will have to be served only a few days later. The referring party should not be tempted to take this first step until he is entirely confident that he can follow through with the next steps. This confidence is best achieved by preparing the referral notice before serving the notice of adjudication. The contents of the referral notice are discussed in Chapter 5, and as we will see in that chapter a substantial amount of information is likely to be required.

If it is thought likely that a request will have to be made to an adjudicator nominating body for the selection of an adjudicator, it is sensible to obtain an appropriate form for the request from the relevant body before serving the notice of adjudication. This will avoid a two day delay later.

Consideration must also be given to the likely procedures that the adjudicator will wish to adopt. This is not easy, because he will have complete discretion about how the adjudication will progress, but if there is a serious dispute about the facts it is quite likely that he will wish to interview relevant witnesses, or arrange a hearing at which the witnesses can be cross-examined. It would be a serious mistake in such circumstances to serve the notice of adjudication shortly before the principal witnesses leave for a fortnight's holiday. Similarly the quantity surveyor who was responsible for the valuation of particular items may be required to explain his valuation to the adjudicator, and his availability should be checked.

One of the great benefits of the adjudication process is the ability to present a case to an independent person for decision during the currency of the work, perhaps before allegedly defective work is removed or covered up. It may be a mistake to delay the start of the procedure until that facility is lost.

On the other hand, the fact that there is no effective time limit for starting an adjudication means that the claimant can take his time in preparing the claim and supporting evidence during prolonged negotiation with the other party. When he is confident that he is ready, he can ambush his opponent with an adjudication reference without any formal prior warning.

Timing is not just a matter for the party who believes he has a claim. Practitioners in the adjudication field joke that the best advice to be given to a client is 'be the claimant', as the party initiating the procedure has initial control of the process and has the opportunity to present his case first to the adjudicator. This advice is more than a joke. If an employer suspects that a contractor may be contemplating starting an adjudication in order to establish an entitlement to an extension of time, he may choose to start the adjudication himself, asking for a decision that there is no such entitlement. This practice is sometimes described as a 'reverse ambush'. It is often used by experienced main contractors in resisting subcontractor claims, and can be very effective. The subcontractor is immediately forced into a defensive position, and often has to present its case before it is ready to do so.

The JCT series of main contracts and the associated Construction Confederation subcontracts (DOM/1 etc.) do not attempt to introduce any limitation on the right to refer disputes to adjudication. They do not use the phrase 'at any time' but neither do they suggest anything to the contrary. The standard form of Government contract GC/Wks/1 goes a little further and states that the employer or the contractor may at any time notify the other of intention to refer a dispute to adjudication.

The ICE 7th Edition form of main contract does however try to impose a restriction by defining when it can be said that a dispute has arisen. Under clause 66(3) it is stated that no matter shall constitute a dispute until a notice of dispute has been served. Such a notice can only be served after the engineer has reviewed the matter in question, or a party to the contract has failed to give effect to a previous adjudicator's decision. Until one of those conditions is satisfied it is said that there is no dispute, but just a 'matter of dissatisfaction'.

Clauses of this type are discussed in Chapter 3. Inclusion of such restrictions renders the whole adjudication procedure as set out in the ICE series of contracts ineffective, because they fail to comply with section 108(2)(a) of the Act. The consequence of such failure is that the Scheme applies, so that not only is the restriction overcome, but the other provisions of the contracts with regard to adjudication are overwritten by the Scheme. This was the conclusion reached by Judge Toulmin when dealing with the Engineering and Construction Contract (otherwise known as the New Engineering Contract) in *John Mowlem plc* v. *Hydra-Tight Ltd* (August 2000).

The Civil Engineering Contractors Association (CECA) standard form of subcontract has a similar restriction. Clause 18 (2) deals with 'submissions' by the subcontractor that a payment is due or that some matter under the subcontract is unsatisfactory. If the contractor thinks that the submission gives rise to a matter of dissatisfaction under the main contract he has an opportunity to pursue that matter with the engineer before it can be called a dispute under the subcontract. Once again it follows that the contract does not satisfy section 108(2)(a) of the Act and that therefore the

Scheme overwrites the adjudication process set out in the sub-contract.

In *R.G. Carter Ltd* v. *Edmund Nuttall Ltd* (Judge Thornton, June 2000), it was held that a requirement to submit to a mandatory mediation procedure before initiating adjudication was unenforceable.

The contract may not set out detailed adjudication provisions at all, but merely incorporate a standard set of adjudication rules published by an institution. Such rules typically remain silent on the question of when a dispute can be referred, thereby impliedly enabling reference 'at any time'. The adjudication rules published by the Technology and Construction Solicitors Association (TeCSA) (formerly the Official Referees' Solicitors Association) go further by stating that notice requiring adjudication may be given at any time and also state that this is not affected by the fact that arbitration or litigation has been commenced in respect of the same dispute.

This provision of the TeCSA rules is in fact unnecessary. There is no suggestion in the legislation or elsewhere that the commencement of other legal proceedings may be a bar to adjudication, and in *Herschel Engineering Ltd* v. *Breen Property Ltd* (April 2000) Sir John Dyson held that there was no reason why an adjudication should not run concurrently with court proceedings. The Act expressly contemplates that the same dispute may form the subject matter of an adjudication and a court action or arbitration, because the adjudicator's decision is only binding until the dispute is finally resolved in such other proceedings, or settled. There is no reason why the other proceedings might not be running at the same time. Breen Property had declined to take any part in the adjudication because there were court proceedings already running, and then unsuccessfully resisted enforcement of the adjudicator's decision.

4.2 *The notice of adjudication*

The first step in the process will be the service of a notice of adjudication. This is little more than a formal statement that one party intends to refer the dispute to adjudication. The requirements regarding the contents of the notice of adjudication vary according to the rules applicable to the contract, the most detailed requirements being those of the Scheme, but the Scheme's requirements are hardly unreasonable. They are set out in paragraph 1(3) of the Scheme:

'(3) The notice of adjudication shall set out briefly –
  (a) the nature and a brief description of the dispute and of the parties involved,
  (b) details of where and when the dispute has arisen,
  (c) the nature of the redress which is sought, and
  (d) the names and addresses of the parties to the contract (including where appropriate, the addresses which the parties have specified for the giving of notices).'

The Scheme does not require any details of the contract itself to be given, but it is useful to include that information, and the ICE Adjudication Procedure requires them.

Clearly the notice of adjudication will not be a substantial document, but it will be of fundamental significance because it defines the 'dispute' that is being referred. If an adjudicator nominating body is to be asked to select an adjudicator it will do so on the basis of the notice of adjudication, and the adjudicator will be nominated to decide the dispute that is identified in the notice. If an adjudicator is named in the contract or there is agreement on who should be appointed, the dispute referred to him will be that described in the notice. He will not be able to decide any other dispute unless the parties agree to widen his jurisdiction under paragraph 8 of the Scheme.

The adjudicator's decision on the matters in dispute will be binding on the parties until the dispute is finally determined by legal proceedings, by arbitration or by agreement (section 108(3) of the Act and paragraph 23 of the Scheme). But the 'matters in dispute' are those identified in the notice of adjudication, and if the adjudicator purports to decide other matters his decision will be of no effect. The position was summarised by Judge Thornton in *Fastrack Contractors Ltd* v. *Morrison Construction Ltd and Impreglio UK Ltd* (January 2000):

'Thus the notice of adjudication; the selection of a person to act as adjudicator by an adjudicator nominating body; the indication from the selected adjudicator of his willingness to act; and the referral notice must all relate to the same pre-existing dispute. Any selection, acceptance of appointment or subsequent adjudication and decision which are not confined to that pre-existing dispute would be undertaken without jurisdiction.'

In *Fastrack*, the defendant main contractor argued that the notice of adjudication had been unsatisfactory in that it raised matters that were not a pre-existing dispute. Prior to delivery of the notice the subcontractor had made an application for payment. The application had included claims in respect of disruption and breach of contract and the like, but there were significant differences between those claims and the claims made in the notice of adjudication. This argument was rejected. Whereas there were differences between the sums claimed, the 'issues had been referred by Fastrack to Morrison, had been rejected by Morrison and therefore ripened into disputes' by the time the notice of adjudication was served. The judge drew a distinction between disputes in which the question is 'what sum is due?' and those in which a specific sum was in dispute.

Simple reference to previous correspondence to define the dispute may not be satisfactory. In *K. & D. Contractors* v. *Midas Homes Ltd* (Judge LLoyd, July 2000), solicitors representing the claimant subcontractor had written to the main contractor's solicitors stating that they were referring a dispute to adjudication under the Scheme. They referred to a series of previous letters and invoices to detail the dispute. The correspondence revealed a variety of possible disputes. The adjudicator had treated all the matters on which the subcontractor relied as being referred properly and made his decision on them in favour of the subcontractor. The judge however decided that only one of the several claims had been covered by the notice of adjudication, and that in dealing with the others the adjudicator had gone outside his jurisdiction. The subcontractor was only able to enforce a part of the decision, and the adjudicator's fees were apportioned between the parties accordingly. The notice of adjudication should have specified the details required by paragraph 1(3) of the Scheme and not just referred to the earlier correspondence.

Real care must therefore be taken to ensure that the 'nature and a brief description of the dispute' are accurately described in the notice. This may not be easy, particularly to the contractor or subcontractor who is not used to expressing matters in contractual terms. If the architect is dissatisfied with a particular part of the works he may instruct that it be removed and rebuilt. The contractor disagrees with the architect's opinion and objects. He expresses the dispute as being whether or not he has to comply with the instruction. The answer may be 'yes', whereas if he described the dispute as being whether or not he is entitled to be paid an extra sum for doing that work a similar answer would have been rather more satisfactory to him.

Oversimplification in describing the dispute is also dangerous. If a dispute has arisen over the value of the works certified on an interim basis, it is tempting to describe the dispute as concerning 'the value of the works' or 'the sum payable in respect of the works' at a particular date. The adjudicator's decision will then be one sum of money, being the total of the valuation or the sum payable. That will be binding on the parties at least for the time being, but it will have no effect on the calculation of the value of the works or the sum payable one month later. If however the disgruntled contractor had also asked whether particular instructions had constituted variations entitling him to additional payment, he might have obtained a decision that would have been of continuing benefit. This problem is not avoided by requesting reasons for the decision. Such reasons would explain how the adjudicator has come to his lump sum decision but would not be binding on the parties. Only the decision itself has that effect.

The other requirements of the Scheme for a notice of adjudication are less sensitive but it is helpful to set them out clearly. It is likely that the notice will be the principal document given to the adjudicator nominating body when it is asked to select an adjudicator. The officer of the organisation responsible for selection, who may be working under considerable pressure to achieve selection of several such adjudicators in a very short time, needs to have all the relevant facts easily available. Specialist knowledge may be an advantage, and location may be important. Conflicts of interest have to be avoided, and it must be made as easy as possible to communicate with all parties involved.

There is no standard form that must be used in order to give an effective notice of adjudication, but the example overleaf is suggested as a form that complies with the requirements of the Scheme. It is set out as a 'legal' document; this is intentional so that it is clear to the recipient that it has a formal purpose. The information required by the Scheme is immediately apparent and is therefore helpful to the adjudicator nominating body (if required) and indeed to the adjudicator who may not receive the referral notice, with full information about the case being presented, until several days after his appointment.

A notice prepared on the basis of this form will satisfy the requirements of all other contractual provisions for the commencement of adjudication, and indeed provides more information than is strictly required by several other sets of rules. The JCT contracts and related subcontracts merely require that the dispute should be 'briefly identified' in the notice. The TeCSA Adjudication Rules require that the notice should identify the dispute in general

Example of notice of adjudication

IN THE MATTER OF THE HOUSING GRANTS, CONSTRUCTION AND REGENERATION ACT 1996[1]
AND IN THE MATTER OF THE SCHEME FOR CONSTRUCTION CONTRACTS[1]
AND IN THE MATTER OF AN ADJUDICATION
BETWEEN
OPTIMISTIC LIMITED
and
MEGABUILD LIMITED

To Megabuild Limited and Complacent Parent plc[2]

TAKE NOTICE that the above-named Optimistic Limited intends to refer the dispute of which particulars are herein set out to adjudication

(1) The nature and brief description of the dispute and relevant contract are as follows:
    1.1 The sum due from Megabuild Limited to Optimistic Limited pursuant to a subcontract for the supply and installation of suspended ceilings at Unit 3, Montezuma Technology Park, Bristol ("the Contract") and/or as damages for breach thereof including:
        (a) The value of its works
        (b) The amount of direct loss and expense alternatively damages incurred by Optimistic Limited as a result of delay and disruption to its works
        (c) Entitlement to payment of £25,000 withheld by Megabuild Limited on the grounds of alleged delay by Optimistic Limited
    1.2 The contract was made on or about 1 April 2000 and incorporated the standard subcontract conditions of Megabuild Limited[3]

(2) The parties involved in the dispute are as follows:
    2.1 Optimistic Limited (suspended ceiling subcontractor)
    2.2 Megabuild Limited (main contractor)

(3) The dispute arose at Bristol on or about 25 July 2000

(4) Optimistic Limited seeks redress of the following nature:
    4.1 An extension of time for completion of the subcontract works from 31 May 2000 to 30 June 2000
    4.2 Payment of £100,000 or such sum as properly represents the sum due to it pursuant to the contract and/or damages, including £25,000 being the sum withheld by Megabuild Limited from the payment made on 25 July 2000 on the grounds of alleged delay.
    4.3 Interest pursuant to the subcontract.

(5) The names and addresses and, if appropriate, the addresses for service of the parties to the relevant contract are as follows:
    5.1 Optimistic Limited
           111 Brightside Lane
           Bristol BS99 7XX
    5.2 Megabuild Limited
           222 Gloomy Park
           Bristol BS99 8XX
    5.3 Complacent Parent PLC[2]
           333 Somewhere Else Street,
           Grimethorpe GR99 9XX

DATED 26 July 2000

                             ....
                             V. Cross, Director
                             for and on behalf of Optimistic Limited

Notes (not forming part of the notice):
[1] If the notice is given under express terms of the contract providing for adjudication, and therefore not under the Scheme, these lines should be omitted.
[2] The notice must set out *all* the parties to the contract, which may include a company giving a parent company guarantee or having some other involvement not apparently relevant to the dispute.
[3] Not a requirement of the Scheme

---

terms. The CIC Model Adjudication Procedure states that the notice should include a brief statement of the issue(s) which are to be referred. GC/Wks/1 does not set out any requirements at all. The ICE Adjudication Procedure states that the notice of adjudication should include the details and date of the relevant contract, the issues that the adjudicator is being asked to decide and the details of the nature and extent of the redress sought, all of which will be found in the example set out opposite.

Failure to comply in every particular with the requirements of the Scheme or relevant contractual rules regarding the form of the notice of adjudication will not necessarily invalidate the notice. An adjudicator nominating body may take the view that no proper notice has been given, but this would be remedied quickly by the service of a new notice. If the designated or nominated adjudicator accepts appointment on the basis of a notice that lacks some required particulars he will normally require the defect to be rectified promptly. It is unlikely that a court would decline to enforce an adjudicator's decision on the sole basis that the notice of adjudication was formally deficient, providing that the details of the dispute being referred are dealt with properly.

4.3 Service of the notice of adjudication

Paragraph 1(2) of the Scheme provides that:

'The notice of adjudication shall be given to every other party to the contract'

No method of service is set out in the Scheme, but it is possible that the contract itself will contain a relevant provision regarding service of documents. If so, that provision should be followed.

If there is no such provision, section 115(2)–(4) of the Act applies:

'(2) If or to the extent that there is no such agreement the following provisions apply.

(3) A notice or other document may be served on a person by any effective means.

(4) If a notice or other document is addressed, pre-paid and delivered by post –
   (a) to the addressee's last known principal residence or, if he is or has been carrying on a trade, profession or business, his last known principal business address, or
   (b) where the addressee is a body corporate, to the body's registered or principal office,
it shall be treated as effectively served'

The party wishing to start adjudication proceedings therefore has a wide choice if the contract is silent as to service. Hand delivery, facsimile or email will all be satisfactory so long as it can be demonstrated that the document was received. Postal service in accordance with section 115(4) will be assumed to have been effective without proof of delivery. In practice, most such notices are served both by facsimile and first class post.

The Scheme requires notice to be given to every other party to the contract. Multi-party construction contracts are uncommon but are found occasionally. A parent company may enter into a contract in order to guarantee the performance of the contractor, or payment by the employer. A finance company may be a party to the contract in addition to the employer. Under the Scheme such additional parties must be identified in the notice of adjudication and served with a copy of it. Other sets of standard adjudication rules do not take account of the possibility of multi-party contracts, typically referring to 'either party' to the contract, and service being effected on 'the other party'. In most circumstances it would be sensible to serve a copy of the notice of adjudication on any third party to the contract, making it clear if appropriate that the third party will not be involved as a party to the adjudication.

The Construction Industry Council (CIC) Model Adjudication Procedure requires the referring party to send a copy of the notice of adjudication to the Adjudicator, if he is identified in the contract.

4.4 Identification or selection of the adjudicator

The notice of adjudication having been served, the clock is ticking. The Act's timetable requires the notice of referral to be delivered to the adjudicator within seven days of the notice of adjudication. The referring party must find an adjudicator without delay. He may manage to persuade the other party to agree that a particular person should be appointed, but given the timescale involved he has little time for discussion. This is unfortunate, because it is often preferable to appoint someone in whom both parties have confidence rather than rely on a person named in the contract, who was probably the choice of one party and therefore subject to some suspicion, however irrational, on the part of the other, or an appointment by a nominating body, in whom perhaps neither party will have confidence.

No particular qualification is required in order to be appointed validly as an adjudicator. The adjudicator nominating bodies may require potential adjudicators to undergo formal training as adjudicators and perhaps to have some other formal qualification as well, but the lack of such qualifications will not affect the validity of the decision. The Scheme says only this:

'4 Any person requested or selected to act as adjudicator in accordance with paragraphs 2, 4 or 6 shall be a natural person acting in his personal capacity. A person requested or selected to act as an adjudicator shall not be an employee of any of the parties to the dispute and shall declare any interest, financial or otherwise, in any matter relating to the dispute.'

Accordingly the adjudicator can be closely involved with one of the parties to the dispute – perhaps regularly retained as a consultant or an employee of an associated company – and still be appointed. Nevertheless he must act impartially (paragraph 12), and he would be well advised to make any connection well known in case it is later argued that he had some interest such as that described in paragraph 4 which he failed to declare.

It is not possible for a professional firm or a company to be appointed as an adjudicator.

Assuming that it has not been possible to agree the appointment of an individual to deal with the dispute, the first place the referring party will look to find the adjudicator is of course the contract. This is made clear in the Scheme:

> '2–(1) Following the giving of a notice of adjudication and subject to any agreement between the parties to the dispute as to who shall act as adjudicator –
> (a) the referring party shall request the person (if any) specified in the contract to act as adjudicator,'

It may seem unlikely that the parties will have specified an adjudicator in the contract, when they have failed to incorporate a system of adjudication that complies with the requirements of the Act. It is however possible, particularly if the contract is on an ICE or similar form which contains a non-compliant system. The adjudication provisions will have been overwritten by the Scheme, but the agreement of a named adjudicator in the contract will still be effective.

It is of course possible that the named adjudicator will be unable or unwilling to act. He may not be able to devote sufficient time to the dispute over the following 28 days, or may be ill or otherwise incapacitated. Subparagraphs 2(1)(b) and (c) deal with such a problem when the adjudicator has already given notice of his non-availability, and also the position where no adjudicator is named:

> '(b) if no person is named in the contract or the person named has already indicated that he is unable or unwilling to act, and the contract provides for a specified nominating body to select a person, the referring party shall request the nominating body named in the contract to select a person to act as adjudicator, or
> (c) where neither paragraph (a) or (b) applies, or where the person referred to in (a) has already indicated that he is

unwilling or unable to act and (b) does not apply, the referring party shall request an adjudicator nominating body to select a person to act as adjudicator.'

Subparagraph (c) deals with the position, perhaps the most common of all, in which there is no adjudicator named in the contract, and no adjudicator nominating body either. The referring party has complete freedom in his choice of adjudicator nominating body, but there is surprisingly little direction as to where he can find such a body. Paragraph 2(3) explains:

'(3) In this paragraph ... an "adjudicator nominating body" shall mean a body (not being a natural person and not being a party to the dispute) which holds itself out publicly as a body which will select an adjudicator when requested to do so by a referring party.'

There are several adjudicator nominating bodies to whom the referring party can turn when none is specified in the contract. The CIArb, the RICS, the ICE and several other professional bodies in the industry maintain panels of adjudicators and will be pleased to provide a name on payment of a fee. The charge at the time of writing by the RICS is £275 including VAT, and the fee charged by the CIArb is £225 plus VAT (£264.38). The subject matter of the dispute does not need to be within the particular discipline of the institution – the RICS for example would be prepared to appoint an adjudicator from its panel to deal with a dispute between a property developer and an architect or between a consultant engineer and a subconsultant.

If however the contract does specify an adjudicator nominating body, that provision will determine to whom application should be made. As with a named adjudicator, the provision will be effective even if the contractual system of adjudication has been overwritten by the Scheme because it does not comply with the requirements of the Act.

The JCT series of contracts and the related subcontracts impose a restriction on the process in that they provide that no adjudicator shall be appointed who will not sign the JCT Adjudication Agreement (discussed in Chapter 6).

The standard Government contract, GC/Wks/1, provides for the appointment of the adjudicator in advance. The Abstract of Particulars appended to the contract gives the name of the adjudicator and of a substitute adjudicator. The contract sets out an agreement to be entered into by the parties and the named adjudicator at an early stage in the contract, and the adjudicator will then be obliged to act in any future dispute unless he is unable to do so because of facts or circumstances beyond his control. The named substitute adjudicator would then be required to act.

4.5 *Request to an adjudicator nominating body*

If there is no adjudicator identified in the contract, and the parties have been unable to agree who should be appointed, the referring party will apply to an adjudicator nominating body for the selection of an adjudicator. Under paragraph 2(1) of the Scheme it is clear that the request can only be made after notice of adjudication has been given. As the timetable is so short, the request will normally be made very shortly after the notice is given, often on the same day. The CIArb suggests in the guidance notes that accompany the application form, that the form should be sent to the Institute with the appointment fee at the same time that the notice of adjudication is given, but in practice this must be read as meaning 'as soon as possible thereafter'.

If the referring party is well prepared, he will have obtained the appropriate form from the adjudicator nominating body named in the contract, or if none is named the body that he has chosen, before the notice of adjudication was delivered. The form can therefore be completed and sent with the fee immediately.

The forms from each adjudicator nominating body vary. The RICS form asks for:

(1) Communication details of the parties and their representatives
(2) Full details of the matters to be adjudicated upon together with a copy of the notice of adjudication
(3) An indication of any special qualifications that the claimant thinks will be required of the adjudicator
(4) A note of any person who it is thought should not be considered for the appointment, perhaps because of a conflict of interest
(5) A copy of the adjudication provisions in the contract, if any.

The CIArb form asks the claimant to provide rather less information, but still requires a copy of the notice of adjudication. Attached to the form however is a comprehensive 'tick-box' list in which the claimant is asked to describe the contract and the subject matter of the dispute by reference to a number of categories.

A request to the Technology and Construction Solicitors Association (TeCSA) or to the Centre for Dispute Resolution (CEDR) should also be accompanied by a copy of the contract or other evidence that the parties have agreed that the TeCSA or CEDR Rules as appropriate apply.

Paragraph 5 of the Scheme requires the adjudicator nominating body to 'communicate the selection of an adjudicator to the referring party within five days of receiving a request to do so'. This can be a challenge for the body concerned. It will try to identify someone on its panel who is suitably qualified to deal with the dispute, whether or not the request form suggests that any specific skill is required. It will also try to avoid selecting an adjudicator who is based in an inconvenient location.

Typically the person at the adjudicator nominating body having responsibility for the selection process will telephone a member of the body's panel and ask if he can accept the appointment. The answer will depend on other commitments and any perception of a conflict of interest, which may have to be researched. The information given in the notice of adjudication and the request to the body to select will be extremely important. The potential adjudicator will try to assess from that information the likely time that he will have to commit to the process to arrive at a decision in 28 days. One (at least) adjudicator nominating body sends a copy of the request by facsimile to a number of panel members, selecting the adjudicator on the basis of the first facsimile to be returned agreeing to accept the appointment.

Having identified an adjudicator who is willing and able to accept appointment, the adjudicator nominating body will write to both parties (normally by facsimile and post) advising them of the name and address of the person selected, and will then have nothing further to do with the matter. This process should have taken no more than five days from receipt of the request to select. Assuming that the request was received by the adjudicator nominating body the day after the notice of adjudication was delivered, there is now perhaps just one day left to deliver the referral notice to the adjudicator.

4.6 Terms of agreement with the adjudicator

The JCT Standard Forms of Contract and the related subcontracts all provide that no adjudicator shall be agreed by the parties or nominated by the adjudicator nominating body named in the contract who will not execute the JCT Standard Agreement for the appointment of an adjudicator. As this is a term of the contract but not a statutory requirement, the parties can of course agree not to bother with the agreement. An adjudicator nominating body who is asked to nominate an adjudicator under a JCT or similar contract will however ask for an assurance from the potential nominee that he is prepared to execute the agreement.

The CIC Model Adjudication Procedure states that the adjudicator is to be appointed on the basis of the CIC's standard form.

The ICE Adjudication Procedure also provides for the execution of a Standard Form of Adjudicator's Agreement. The status of the ICE procedure is in some doubt, as the term of the contract providing for adjudication does not comply with the Act and therefore the provisions of the Scheme overwrite the contractual provisions. The adjudication might therefore proceed without any reference to the ICE procedure, but if the parties choose to use the procedure and they and the adjudicator sign the ICE Adjudicator's Agreement, the relationship between the parties and the adjudicator will be governed by that document.

The standard form of Government contract, GC/Wks/1, envisages that the adjudicator will be appointed in advance of any dispute arising. The Adjudicator's Appointment form therefore deals with future disputes and contemplates that there may be more than one. This form is to be executed as a deed.

The Scheme itself does not say anything specific about the nature of the adjudicator's relationship with the parties, but nevertheless it is clear that it is contractual. The adjudicator cannot be required to act without his agreement, and his appointment can be brought to an end either by resignation or revocation, with financial consequences in both cases (see below). The adjudicator's powers and duties are set out in the Scheme (discussed in Chapters 6 and 7), and the Scheme provides for him to be paid reasonable fees and expenses (discussed in Chapter 8). By requesting him to act, either directly or through an adjudicator nominating body, the referring party is making an offer to him, both on his own behalf and as agent for the other party to the dispute. The authority to act as agent is given to him by the contract. The contract will either contain an express term providing for adjudication, or such a term will be implied by the Act. It is a necessary implication that either party has the authority to appoint the adjudicator on behalf of both parties.

By agreeing to act, the adjudicator is accepting the offer. A contract therefore comes into existence and no further agreement needs to be made. With the exception of the Government form, dealing with future disputes, the standard forms of appointment such as those described above are therefore not essential. They assist in clarifying the basis on which the adjudicator is acting, particularly with regard to fees, but the adjudicator can be validly appointed without the execution of a form.

The terms of the various standard forms will be discussed in later chapters dealing with the relevant procedural points under the Scheme and the equivalent contractual provisions.

4.7 Procedure if the appointment system fails

It is an essential requirement of adjudication that the system can be used effectively at very short notice. The first step is the appointment of the adjudicator, and if that cannot be achieved within a very few days, the whole system will founder. There can be no guarantee that the individual chosen by the parties will be available immediately when required. Similarly there can be no guarantee that the institution or commercial organisation acting as an adjudicator nominating body will deal with the request for a nomination efficiently. The Scheme therefore provides a system of alternative fail-safe procedures.

If an adjudicator is named in the contract, the first approach will be to him. Under paragraph 2(2), he has two days from receipt of the request in which to reply:

> '2(2) A person requested to act as adjudicator in accordance with the provisions of paragraph (1) shall indicate whether or not he is willing to act within two days of receiving the request.'

The potential adjudicator is not under any contractual or other obligation to comply with this, unless he has previously entered into some other agreement with the parties to do so. The request is no more than an offer to appoint him. The significance of the two day period is that if he fails to reply within that period, the referring party can turn to paragraph 6(1), which provides:

> '6-(1) Where an adjudicator who is named in the contract indicates to the parties that he is unable or unwilling to act, or where he fails to respond in accordance with paragraph 2(2), the referring party may –
> (a) request another person (if any) specified in the contract to act as adjudicator, or

(b) request the nominating body (if any) referred to in the contract to select a person to act as adjudicator, or
(c) request any other adjudicator nominating body to select a person to act as adjudicator.'

The same procedure applies if the other person mentioned in paragraph 6(1)(a) fails to respond. The non-availability of a named adjudicator is therefore covered. If the first named person is not available, the referring party has a wide range of options. He can turn to another named person if there is one, but he does not have to do so. He can go to any adjudicator nominating body instead.

If a request is made to an adjudicator nominating body, that body has five days to respond with a selection:

'5-(1) The nominating body referred to in paragraphs 2(1)(b) and 6(1)(b) or the adjudicator nominating body referred to in paragraphs 2(1)(c), 5(2)(b) and 6(1)(c) must communicate the selection of an adjudicator to the referring party within five days of receiving a request to do so.'

Once again, there is no contractual obligation on the adjudicator nominating body imposed by the Scheme, although there may be such an obligation through normal contractual principles, a fee having been paid for the service. The point of the time limit is to enable the referring party to go elsewhere if the adjudicator nominating body fails:

'5-(2) Where the nominating body or the adjudicator nominating body fails to comply with paragraph (1), the referring party may –
(a) agree with the other party to the dispute to request a specified person to act as adjudicator, or
(b) request any other adjudicator nominating body to select a person to act as adjudicator.'

There is then a potentially endless succession of new possibilities if the alternatives fail. There is a recurring right to go elsewhere if the chosen adjudicator does not accept the appointment or the adjudicator nominating body does not select an adjudicator. If these fail-safe measures are used it is highly unlikely that the appointment will be achieved in time to allow the referral notice to be delivered to the adjudicator within seven days from the date of the notice of adjudication, as is required under paragraph 7 of the Scheme. This will not however affect the validity of the process.

4.8 Objections to specific adjudicator

If the adjudicator has been named in the contract, he may effectively have been chosen by one of the parties and imposed on the other. This may not seem significant at the time the contract is signed, but when a dispute arises the claiming party may feel that he would prefer someone appointed independently. If on the other hand the adjudicator has been selected by an adjudicator nominating body either or both parties may have doubts about the suitability of the person nominated.

Mere objection to the appointment has no effect on the process or on the validity of the decision. Paragraph 10 provides:

'10. Where any party to the dispute objects to the appointment of a particular person as adjudicator, that objection shall not invalidate the adjudicator's appointment nor any decision he may reach in accordance with paragraph 20.'

The appointment must however be in accordance with the terms of the relevant contract. If the Scheme is to be applied to the contract, either because the contract was silent as to adjudication or because it did not comply with the requirements of the Act, the limitation on the potential persons to be appointed is contained within paragraph 4:

'4. Any person requested or selected to act as adjudicator in accordance with paragraphs 2, 5 or 6 shall be a natural person acting in his personal capacity. A person requested or selected to act as adjudicator shall not be an employee of any of the parties to the dispute and shall declare any interest, financial or otherwise, in any matter relating to the dispute.'

Under the Scheme, therefore, the adjudicator cannot be a limited company or a firm, nor can he be an employee of either party. There is however nothing to stop a partner in a firm of consultants regularly retained by one of the parties from being named in the contract and subsequently appointed. It is even theoretically possible for the contract to name the architect as the adjudicator, although he would have to declare an interest in any dispute that was affected by his role as architect. There may be serious doubts about his ability to act impartially as he is required to do by paragraph 12 (to be discussed in Chapter 6), but his appointment would not be invalid.

The standard form contractual systems vary in their required qualifications for the potential adjudicator. The JCT contracts and associated subcontracts do not bar an employee of one of the parties, but do insist that the adjudicator be prepared to execute the standard JCT Adjudication Agreement.

The Government contract GC/Wks/1 requires the adjudicator to be independent of the employer, the contractor, the project manager and the quantity surveyor, and if the appointed adjudicator at any time ceases to be independent a substitute adjudicator is to be appointed. The CIC Model Adjudication Procedure does not exclude any category of potential adjudicator.

If the adjudicator named in the contract is in fact unable to act because he is not qualified under the Scheme or relevant contractual provision, the objecting party should proceed as with any other objection to jurisdiction (discussed in Chapter 5).

Adjudicator nominating bodies take real care to select suitable adjudicators but can sometimes make surprising choices. A well-known example occurred early in the operation of the process, when an adjudicator nominating body selected a quantity surveyor based in Wigan to adjudicate in a dispute over £8000 which required a site visit in Somerset. The referring party may request a particular skill or discipline and be disappointed when he finds that the person selected does not have that quality. Nevertheless such a selection under any of the standard forms is perfectly valid. If the person selected accepts the offer of an appointment, he is validly appointed and cannot be removed on the insistence of either party.

If both parties agree that the selected person is not suitable to adjudicate in the dispute, they can revoke his appointment and either appoint an agreed person or make another application to an adjudicator nominating body in the hope that a more appropriate person is selected.

4.9 *Revocation of appointment and resignation of the adjudicator*

The relationship between the parties and adjudicator is contractual, and contracts can of course be brought to an end. Paragraph 11 of the Scheme deals with revocation by the parties:

'11–(1) The parties to a dispute may at any time agree to revoke the appointment of the adjudicator. The adjudicator shall be entitled to the payment of such reasonable amount as he may determine by way of fees and expenses incurred by him. The parties shall be jointly and severally liable for any sum which remains outstanding following the making of any determination on how payment shall be apportioned.

(2) Where the revocation of the appointment of the adjudicator is due to the default or misconduct of the adjudicator, the parties shall not be liable to pay the adjudicator's fees and expenses.'

If the parties settle the dispute after the adjudicator has been appointed, they will not require him to do any more work. They will not ask him to prepare a 'consent decision' in the way that they might request a consent award in an arbitration. A consent decision would not be a decision at all, and there would be no advantage in having such a document. It would be of no legal effect, whereas a consent award in arbitration is an arbitrator's award and can be enforced as such. The parties will simply ask the adjudicator to stop work, effectively revoking his appointment under paragraph 11.

The main provision for payment of fees and expenses is found in paragraph 25 and is only effective upon the making of a decision by the adjudicator, so paragraph 11 contains a further provision entitling him to payment. As under paragraph 25 the quantification of those fees and expenses, subject to a requirement of reasonableness, is a matter for the adjudicator and the same principles apply (considered in Chapter 8).

The parties may agree to revoke the appointment in other circumstances. They may agree that the adjudicator is not suitably qualified to deal with the particular matters in dispute, or they may decide not to use the adjudication procedure at all, preferring to go immediately to some other method of dispute resolution. It must be remembered that it is not just up to the referring party, who initiated the procedure. If for example the referring party realises that the other party is insolvent and will not be able to comply with a decision that he pay, it may be the referring party's wish that the adjudicator stop working and incurring cost. The adjudicator cannot stop work on that basis, but must have the agreement of both parties that his appointment should be revoked.

The third sentence of paragraph 11(1) is difficult to understand. It is suggested that the adjudicator may make some determination as to how his fees and expenses should be apportioned between the parties upon revocation. Revocation of the appointment has to be by the agreement of the parties, and in coming to that agreement it would be normal for the parties to decide how the adjudicator's fees are to be split between them. It is difficult to imagine circumstances in which the parties agree that the dispute is to be settled, or the adjudication process terminated for some other reason, without agreeing how the fees are to be apportioned. Nevertheless if they should do so, it seems that the adjudicator can deal with that question, and if he does not do so the fees are the joint and several responsibility of both parties.

Paragraph 11(2) of the Scheme is also difficult to understand in a practical context. There is no provision for dismissal of the adjudicator for misconduct. Assuming therefore that the misconduct is not sufficiently serious to amount to repudiation, the parties will still have to revoke the appointment under paragraph 11(1) if they want to terminate the relationship with the adjudicator. They will have to agree on such a course of action. If they wish to dispute an obligation to pay the adjudicator's fees and expenses they will have to show that the reason that they revoked the appointment was the adjudicator's default or misconduct of the adjudication.

For the adjudicator to be in default, he must be failing to carry out his obligations. Those obligations are also the mandatory provisions for the conduct of the adjudication, and failure to observe them would seem to be misconduct. It is difficult to see what else might be covered by the phrase 'misconduct of the adjudication'. The obligations are:

(1) To act impartially (paragraph 12(a))
(2) To act in accordance with any relevant terms of the contract (paragraph 12(a))
(3) To reach a decision in accordance with the applicable law in relation to the contract (paragraph 12(a))
(4) To consider any relevant information submitted to him (paragraph 17)
(5) To make available to the parties any information to be taken into account in reaching his decision (paragraph 17)
(6) To reach a decision within the prescribed time limits (paragraph 19)
(7) To provide reasons for his decision, if requested by one of the parties (paragraph 22).

It is unlikely that the parties will agree to revoke the appointment of the adjudicator because of his default in most of the above obligations, as any such default will be for the benefit of one of them. The most likely reason for such an agreement will be the failure of the adjudicator to reach his decision in time. As will be seen in Chapter 6, either party has the right to appoint a replacement adjudicator in such circumstances, in which case the parties are likely to agree that the first appointment should be revoked and no payment made.

Any failure to comply with the last of the obligations listed above, to provide reasons if asked to do so, will become apparent when the decision itself is delivered. The adjudicator's appointment will then be terminated anyway, and it will be too late to revoke it. The decision will not be invalid, and the appropriate action for the party who wanted the reasons will be to seek an order of the Court that the adjudicator give the reasons in accordance with his obligation.

The appointment may be terminated on the initiative of the adjudicator. Paragraph 9 provides:

> '9–(1) An adjudicator may resign at any time on giving notice in writing to the parties to the dispute.
>
> (2) An adjudicator must resign where the dispute is the same or substantially the same as one which has previously been referred to adjudication, and a decision has been made in that adjudication.
>
> (3) Where an adjudicator ceases to act under paragraph 9(1) –
>   (a) the referring party may serve a fresh notice under paragraph 1 and shall request an adjudicator to act in accordance with paragraphs 2 to 7; and
>   (b) if requested by the new adjudicator and insofar as it is reasonably practicable, the parties shall supply him with copies of all documents which they had made available to the previous adjudicator.
>
> (4) Where an adjudicator resigns in the circumstances referred to in paragraph (2), or where a dispute varies significantly from the dispute referred to him in the referral notice and for that reason he is not competent to decide it, the adjudicator shall be entitled to the payment of such reasonable amount as he may determine by way of fees and expenses reasonably incurred by him. The parties shall be jointly and severally liable for any sum which remains outstanding following the making of any determination on how the payment shall be apportioned.'

Paragraph 9(1) makes it clear that an adjudicator acting under the Scheme will not be in breach of contract if he chooses to resign. He does not have any right to payment for work done to the date of resignation, unless the circumstances of the resignation are covered by paragraph 9(4).

The adjudicator has no option but to resign if it becomes clear that the dispute has already been dealt with in an adjudication. The previous decision is after all contractually binding, and there is nothing further that the adjudicator can do. Subtle changes in the way that the argument is expressed will not be sufficient to avoid the effect of this clause.

This clause of the Scheme was the subject of a challenge to enforcement of a decision in *Sherwood & Casson Ltd* v. *Mackenzie Engineering Ltd* (November 1999, Judge Thornton). An interim account had been referred to adjudication by the contractor and a decision had been given about the value of variations. Subsequently a final account was prepared which included a loss and expense claim. A further dispute arose about the amount payable. The dispute included the valuation of variations, some of which had been in the adjudication about the interim account. The judge decided that on the facts of this case the adjudicator was justified in finding that the final account exercise was sufficiently different from the interim valuation process to mean that the two disputes were not substantially the same. He accepted the adjudicator's view expressed in his decision that the valuation of variations on an interim basis, when the loss and expense claims were not being considered, was different to the exercise of valuing the same variations in the context of the final account when such claims were in consideration. Judge Thornton commented that the adjudicator does have some power under paragraph 9(2) to decide for himself whether the two disputes are in fact substantially the same and the court would not be quick to overrule him:

'In conducting that enquiry, the court would give considerable weight to the decision of the adjudicator and would only embark on a jurisdictional enquiry in the first place where there were substantial grounds for concluding that the adjudicator had erred in concluding that there was no substantial overlap.'

If an adjudicator is obliged to resign under paragraph 9(3) he will be entitled to payment for work done and can determine how responsibility for that payment is to be apportioned.

Paragraph 9(4) also gives the adjudicator the right to payment if he resigns because he finds that he is being asked to deal with a dispute that is significantly different to that referred to him in the referral notice and for that reason he is not competent to decide it. This is not a helpful provision. The adjudicator will have been appointed to deal with the dispute contained in the notice of adjudication. If the dispute described in the notice of referral is significantly different from that, the adjudicator will not have jurisdiction to deal with it. This problem is considered in Chapter 5. The Scheme gives the adjudicator the right to be paid if he resigns in those circumstances.

Having taken on the task of adjudicating the dispute set out in the notice of adjudication, it is difficult to see why the adjudicator should have special rights on resignation because the dispute turns out to be different to that described in a document that is delivered several days later than that notice. It may be that the notice of adjudication was properly prepared and accurately describes the matters in dispute, whereas the referral notice seriously misdescribes the dispute.

Whatever the reason for the resignation, save as set out in paragraph 9(2) that the matter has already been adjudicated, the referring party has no option but to start all over again with a new notice of adjudication and appointment procedure.

The Scheme does not contemplate the possibility that the adjudicator may die or otherwise become incapable of acting in the adjudication. Such a misfortune will of course lead to the failure of the adjudicator to produce his decision within the relevant time limit, which will enable the parties to restart the process, and the parties can agree to revoke the first appointment.

The JCT, in drafting the adjudication provisions for their contracts, were less satisfied that adjudicators would be immortal, and provided that if the adjudicator dies or becomes ill or is unavailable to act, the parties can agree on a replacement or can apply to the agreed nominating body. There is no provision entitling the adjudicator to resign.

The CIC Model Procedure also contains a term that if the adjudicator is unable to act or fails to reach his decision within the time required, either party can request the nominating body to nominate a replacement. The CIC also enable the adjudicator to resign, but the procedure says nothing about payment of fees in such circumstances. The position would seem to be that he would not be entitled to any fee as he would not have produced a decision.

The TeCSA Rules give the chairman of the Association the power to replace the adjudicator when it appears necessary to do so. He will consider whether this is the case if any party represents to him that the adjudicator is not acting impartially, is physically or mentally incapable of conducting the adjudication or is not dealing with matters 'with necessary dispatch'. There is no provision for resignation by the adjudicator.

Unlike the Scheme, these institutional rules do not contain any mandatory requirement for the adjudicator to resign if he finds that the dispute is the same or substantially the same as a dispute previously referred to adjudication. Nevertheless he will be faced with the problem that the decision in the previous adjudication will be binding until the dispute is finally determined by legal proceedings or arbitration.

CHAPTER FIVE
PRELIMINARY MATTERS – THE REFERRAL NOTICE AND JURISDICTION

5.1 Time for delivery of the referral notice

The notice of adjudication was the document that started the whole adjudication process. The adjudicator has been appointed to deal with the dispute or disputes described in the notice of adjudication, but that notice will have contained few details. The adjudicator cannot begin the process properly until a detailed statement of the referring party's case has been delivered. The referral notice is that document.

Section 108(2)(b) of the Act requires the contract to provide a timetable with the object of securing the appointment of the adjudicator and the referral of the dispute to him within seven days of the service of the notice of adjudication. It does not require the referral notice to be delivered within that seven day period, but merely requires a timetable to be in place with the stated objective.

The Scheme's procedures for appointment of the adjudicator, considered in Chapter 4, are designed to work in accordance with this timetable:

- Day 1   Notice of adjudication given to other party and request made to an adjudicator nominating body for the selection of an adjudicator
- Day 4/5   Adjudicator nominating body advises the referring party of the identity of the person selected by it to be the adjudicator
- Day 7   The person selected indicates that he is prepared to accept appointment as adjudicator

There is nothing in the Scheme to suggest that weekends or even bank holidays do not count in this timetable. Seven days means seven days.

Whilst the timetable satisfies the requirements of the Act in that it has the object of securing the appointment within seven days, the process may well take a little longer. The referring party may not have delivered the request for appointment to an adjudicator nominating body on the same day as the notice of adjudication is delivered, and the five days that an adjudicator nominating body is allowed to select an adjudicator may therefore not start to run on day 1. The appointment procedure may encounter problems requiring the fail-safe procedures to be used. It is not unusual for the adjudicator not to be appointed within the seven day period envisaged by the Act and the Scheme.

The Scheme's timetable continues to deal with the date for delivery of the referral notice:

'7–(1) Where an adjudicator has been selected in accordance with paragraphs 2, 5 or 6, the referring party shall, not later than seven days from the date of the notice of adjudication, refer the dispute in writing ("the referral notice") to the adjudicator.'

As the summarised timetable above shows, the identity of the adjudicator may not be known, even if the procedures have been running to time, until the seventh day. If that is the case the referring party is supposed to deliver the referral notice to him on the same day. If the procedures have not run to time, it will not be possible even in theory to comply with the Scheme's requirements.

A strict reading of paragraph 7(1) suggests that a failure to meet the seven day deadline is fatal to an adjudication under the Scheme, because apparently the referral notice cannot be delivered after the seventh day. Adjudicators are unlikely to be too worried about this problem provided that the referring party is not delaying matters unnecessarily. A delay at this stage is not eating into the period allowed for the adjudicator to reach his decision which only starts to run on receipt of the referral, and it would be unhelpful to the parties to decline to deal with the dispute because the referral notice is a few days late.

A more serious problem may arise if the responding party decides to take a jurisdiction point as a result of the delay. Such issues are considered later in this chapter. Again it is unlikely that the courts will wish to undermine the process of adjudication by applying a strict interpretation to this timetable provision. The strict approach might be consistent with the words of paragraph 7(1), but would mean that the fail-safe procedures, designed to ensure that an adjudicator can be appointed eventually despite repeated practical problems, would always be a complete waste of time. This

Preliminary Matters – The Referral Notice and Jurisdiction 85

cannot have been the intention of those responsible for drafting the Scheme.

The referring party should try to meet the seven day deadline for delivery of the referral notice if at all possible, which means that the document must be prepared well in advance and preferably before the notice of adjudication is served. If however a few days' delay is unavoidable, an adjudication under the Scheme should be able to go ahead.

Adjudication is a contractual process, and it cannot be assumed that the position under the Scheme is also the position under other contractual sets of adjudication rules. The amendments to the standard forms of JCT contract published in April 1998 gave a slightly different timetable for the delivery of the referral notice. The referral notice was to be delivered within seven days from the date of the notice of adjudication, or, if later, within seven days from the execution of the JCT Adjudication Agreement by the adjudicator. As the adjudicator has of course first to be selected, this was bound to add several days to the timetable. By the time the JCT amendments had been consolidated into the 1998 standard forms, this had changed, and the date of execution of the adjudication agreement was no longer relevant.

The position under the JCT contracts and associated subcontracts is now that the referral, as the referral notice is described in these contracts, is to be given to the adjudicator within seven days of the notice of intention to refer (the notice of adjudication), if the adjudicator has been appointed within that time. If he has not been appointed within seven days, it must be given to him immediately on appointment. The position is therefore the same as under the Scheme, except that the JCT have expressly dealt with the problem of delay in appointment.

Because the ICE standard forms do not comply with the Act in allowing parties to refer disputes to adjudication at any time, the Scheme applies to all ICE contracts. Nevertheless the provisions of the ICE Adjudication Procedure will be contractually effective insofar as they do not conflict with the Scheme. Under paragraph 4.1 of that procedure, a full statement of the referring party's case must be sent to the adjudicator and the other party within two days of receipt of confirmation of the adjudicator's appointment. This should also still be within seven days of the date of the notice of adjudication in order to comply with the Scheme.

The Construction Industry Council Model Adjudication Procedure has a more practical approach that avoids the potential problems of strict interpretation of the Scheme. Under paragraph 14 of the procedure, the referring party is required to send a full statement of case to the adjudicator within seven days of the notice of adjudication, or as soon thereafter as the adjudicator is appointed.

Under GC/Wks/1 the adjudicator is already identified in the contract, and assuming that he is able to act it should be easier to comply with the seven day timescale, which is imposed under this contract in a similar manner to the Scheme.

The Technology and Construction Solicitors Association adopts a quite different approach. Whilst the other systems, including the Scheme, require the referral of the dispute to the adjudicator to be achieved by delivery of the referral notice or similar document, the TeCSA Rules define 'referral' as acceptance of appointment. The date of referral is therefore the date that he confirms his acceptance. Their timetable seeks to achieve the objective required by section 108(2)(b) of the Act without any referral notice at all. The delivery of a full statement of case then becomes something for the adjudicator to order or not as he thinks appropriate. It is difficult to envisage an adjudicator proceeding without such a document, but it is not impossible. There is no timetable for it within the rules.

5.2 Form and contents of the referral notice

The Scheme says little about the referral notice. Having dealt in paragraph 7(1) with the timing of it, the Scheme goes on to say what should accompany it and that it should be copied to the other parties:

> '7-(2) A referral notice shall be accompanied by copies of, or relevant extracts from, the construction contract and such other documents as the referring party intends to rely upon.
>
> (3) The referring party shall, at the same time as he sends to the adjudicator the documents referred to in paragraphs (1) and (2), send copies of those documents to every other party to the dispute.'

This does not help the referring party to prepare its referral notice. In practice, the referring party will wish to set out its case fully but clearly. These two objectives may be difficult to reconcile. In preparing it, some thought should be given to the person to whom it is addressed, the adjudicator. He has 28 days from receipt of the referral notice in which to reach his decision. He needs to be able to understand the issues immediately. He also needs to have all the relevant facts, and if there are points of law involved he needs to have them explained and argued. Several files of supporting papers may be necessary to achieve these objectives, but they will not be greeted with enthusiasm. Ideally the merits of the referring party's case should seem obvious. As with every good magician's trick, a great deal of work may be necessary to achieve the required simplicity.

Lawyers may assume that they are best qualified to prepare referral notices because they have so much experience of preparing pleadings in court cases and arbitrations. Non-lawyers who have been unfortunate enough to have to read substantial pleadings may not have found them particularly clear and simple, and it should be remembered that many adjudicators are not lawyers. In fact the traditional pleading is not appropriate for most adjudication cases. The court pleading is only a part of the presentation of the party's case to the court. It sets out the relevant facts but does not necessarily explain the legal basis for the claim, that being left to legal argument in court. It seldom anticipates the other party's case, expecting that the other side will serve its pleading which will then be followed by a reply. It is not accompanied by evidence.

The referral notice in an adjudication may be the only document that the referring party will put before the adjudicator. It needs to leave the adjudicator with the impression that the referring party's case seems obviously right, so that the other party has to fight just to regain the starting line. In an appropriate case therefore it may comprise the following.

5.2.1 A simple narrative, explaining what the project is, what work was being done and any relevant circumstances

This will typically be a short introduction to the main claim narrative. It should set the scene, giving the general description of the project, introducing the relevant parties and explaining what their roles were. It should set out any specific aspects that the adjudicator should have in mind, such as the urgency or particular difficulty of the work being done.

5.2.2 What the dispute is about

A traditional pleading often gives no clue about the particular matter in dispute in litigation until the reader has struggled through several pages of formal recital stating how the parties entered into the contract and what the terms of the contract were. The adjudicator should be told about the reason for his appointment before he is asked to consider the terms of the contract. He will then be able to focus on the relevance of specific contract terms. It is appropriate to tell him at an early stage in the referral notice that, for example, the dispute concerns the deduction of contra charges on the basis of alleged defective work, and that the claimant subcontractor not only contests that the work was defective, but also says that insufficient notice of intention to withhold payment has been given. The detail of the case will be in the sections that follow.

5.2.3 A statement of the relevant contractual terms

Adjudication is a contractual process, and the adjudicator will wish to be absolutely certain, if he can, of the contract terms that he must consider. His first consideration will be the terms relating to the adjudication process itself. He will wish to establish, if only in his own mind, that he has jurisdiction. As we will see later in this chapter, adjudicators have been encouraged by comments in several judgments of the Technology and Construction Court to carry out investigations to satisfy themselves that they have jurisdiction, even if their decisions on the point are not binding on the parties. He will also want to know whether there are any express contractual rules for the conduct of the adjudication, or alternatively whether the Scheme for Construction Contracts applies. The following contractual matters will typically need to be set out clearly:

(1) The date of the contract, to show that it falls within the scope of the Act
(2) That it is a contract in writing, within the meaning of the Act, with reference to relevant documents
(3) Whether there is a contractual system of adjudication that complies with the Act's requirements, or alternatively that the Scheme applies
(4) The terms relating to the specific dispute, e.g. the mechanism for payment, the obligations regarding time or quality, the terms relating to valuation of variations or entitlement to loss and expense
(5) Any contractual entitlement to interest on late payment.

The contract documents should be copied and supplied so that the adjudicator can check the statements that are made. It is not always necessary to copy all the contract documents – for example a subcontractor's claim for the value of a variation is unlikely to need consideration of the main contractor's site safety rules, and several pages of the subcontract order documents therefore will not need to be copied.

5.2.4 The specific facts that establish the contractual entitlement

This is the place for the detail of the claim. The adjudicator is unlikely to be impressed by generalised statements of how unreasonable the other party has been. He will require hard facts. On what date was the instruction given, by whom and to whom? Where does the adjudicator find the written evidence? Which witness can support this narrative? How has the evaluation been calculated and on what assumptions? Have the contractual preconditions to entitlement been satisfied? It will be appreciated that this is much more demanding than the requirements of a pleading in a court action. All the 'Requests for Further and Better Particulars' must be anticipated and answered, and all the relevant evidence must be produced.

There is a real danger that the trees may obscure the wood. If the dispute is about a matter of any real complexity, or about a number of matters as in a typical dispute about a final, or even interim, account, real skill may be needed to present the case with clarity. If clarity is not achieved, the advantage of having the initiative as claimant will be lost. The use of appendices setting out details of dayworks or valuation of lists of variations may help, and the provision of such information in spreadsheet form on computer disk may be welcomed by the adjudicator. All relevant correspondence, memoranda, minutes of meetings etc. should be produced in an indexed file, with references in the narrative. Irrelevant documents should not be included in the bundle sent to the adjudicator.

5.2.5 The other party's argument, and why it is wrong

The claimant has a real advantage in being the first to present his case, but the adjudicator will be interested in knowing what the other side is saying. If the respondent has given any explanation of his case in previous correspondence or discussion, the claimant has a further advantage in being able to explain his opponent's position. He can then go on to demonstrate that the case is wrong – hopefully he will be able to suggest that it is clearly wrong and explain why. This has the obvious advantage of setting the adjudicator against the respondent from the start, and if the defence case changes substantially by the time the response document is served, the adjudicator will know that the respondent has abandoned the position that he had adopted before and may be suspicious. There is of course a danger that the claimant will demonstrate that in truth he has completely misunderstood the respondent's position, and so this approach must be used with care.

5.2.6 Any relevant law

The adjudicator may or may not be a lawyer. It is likely that he will have had some legal training, particularly in contract matters, but it should not be assumed that he will be aware of all relevant statute or case law. If the claimant needs to rely on any specific authority it should be stated and a copy produced.

5.2.7 A summary of the decision sought

The referral notice should conclude with a succinct statement of what the claimant asks the adjudicator to order. This is broadly equivalent to the 'prayer' at the end of a traditional pleading. If the claimant seeks a decision that is effectively a declaration of the contractual position, the express words sought should be stated. If a sum of money is claimed it should be calculated clearly, stating whether VAT is claimed in addition. Any alleged entitlement to interest should also be calculated. Although the costs of the adjudication are not normally recoverable, if particular circumstances suggest that such a claim can be made the figure for costs to date should be set out with a statement (if appropriate) that a further claim will be made prior to the making of the decision for costs incurred during the adjudication.

Finally, in considering the content of the referral notice, it must be remembered that the referral notice should be confined to the dispute forming the subject of the adjudication, set out in the notice of adjudication. The adjudication does not concern other matters, and it is important to ensure that the notice of adjudication correctly describes the dispute. Omissions cannot be rectified by the notice of referral, and as we shall see in the next section, there are likely to be real difficulties in expanding the adjudication to deal with matters forgotten in preparing the first notice.

5.3 Related and unrelated disputes

The notice of adjudication, no matter how well drafted, may not encompass all the matters that need to be resolved in order to arrive at a fair and correct decision. A contractor may seek a decision as to the sum of money that should have been paid on a particular date. That may require consideration of whether a notice of intention to withhold payment based on a claim for liquidated and ascertained damages by the employer was given at the appropriate time, and whether or not the contractor was entitled to an extension of time for completion of the works. The ability of the adjudicator to deal with such matters is governed by the rules imposed on him by the express words of the contract or by the implication of the Scheme.

Paragraph 20 of the Scheme reads as follows:

> '20   The adjudicator shall decide the matters in dispute. He may take into account any other matters which the parties to the dispute agree should be within the scope of the adjudication or which are matters under the contract which he considers are necessarily connected with the dispute...'

The first option for widening the scope is self-evident from the contractual nature of the adjudication process. The parties can agree to include any matter, but as the relationship with the adjudicator is itself contractual, he is not obliged to take such matter into account. It may be, for example, that he has neither the necessary skills to consider such wider matters, nor the time or resource to consider them within the 28 period in which he is committed to reach his decision.

He also has a discretion to consider matters about which there is no agreement, if he thinks that they are necessarily connected to those referred to him. In *Northern Developments (Cumbria) Ltd* v. *J. & J. Nichol* (January 2000, Judge Bowsher) the court considered a refusal by an adjudicator to consider a claim for damages for repudiation of contract. The claimant in the adjudication had asked the adjudicator to decide the question of outstanding moneys in respect of work carried out. The respondent said that it was entitled to set-offs in respect of defective work, delays and damages for alleged repudiation. The adjudicator decided that these set-off claims were necessarily connected with the original claim, but also decided that the claim for damages for repudiation was not a matter 'under the contract'. He therefore did not consider it. This was clearly wrong. Repudiation was a matter that arose under the contract, as had been decided in the context of arbitration in 1942 in *Heyman* v. *Darwins*. Nevertheless he had been right not to consider it because the alleged repudiation had occurred after the date for giving notice of intention to withhold from the payment for which the claimant had started the adjudication. Therefore the set-off for damages arising from the repudiation had not been the subject of a valid notice of intention to withhold. He had arrived at the right decision for the wrong reason.

Judge Bowsher dealt with the discretion that is given to the adjudicator in paragraph 20:

'Paragraph 20 says that the adjudicator *may* take such matters into account. If he had the discretion it would be a wrongful exercise of his discretion to refuse to exercise the discretion. If he did exercise such a discretion, it would almost certainly be impossible to challenge the exercise of that discretion whichever way he decided the discretion, in favour of or against considering the other matters.'

Whilst the adjudicator has a discretion to 'take into account' such wider matters, he does not have discretion to make a decision with regard to them. This is an important distinction. If for example the contractor gives notice of adjudication asking for a decision as to the value of his interim account, the adjudicator's decision will deal with that question only. He will no doubt take into account the question as to whether a particular alleged variation was indeed a variation that entitled the contractor to payment. His decision that the contractor is entitled to a sum of money in respect of the interim account, which will incorporate the value that he places on that variation, will be a decision that will be binding until finally resolved in court or in arbitration, but the decision that he made on the question about the variation will not be binding, and may well be a dispute in succeeding valuations of interim payments.

The TeCSA Rules provide that the scope of the adjudication shall include not only the matters identified in the notice of adjudication and further matters which the parties agree should be included, but also any further matters which the adjudicator determines must be included in order that the adjudication may be effective and/or meaningful. This is much wider than the Scheme, and enables the adjudicator to give a binding decision on the questions that he decides should be included.

The CIC Rules do not enable the adjudicator to take the initiative in this way. He is restricted to deciding the matters set out in the notice of adjudication and other matters that the parties and he agree shall be within the scope of the adjudication.

5.4 Questions of jurisdiction

Whilst some assistance is given in deciding how the adjudicator is to deal with related disputes, the Scheme is silent on other matters of jurisdiction and does not give any guidance as to how disputes about jurisdiction should be dealt with. Such disputes can arise when the responding party argues that the contract was not in writing within the definition of the Act, was formed before 1 May 1998 when Part II of the Act came into force, or for some other reason was not a construction contract. He may argue that the contract was one of those exempted from the operation of the Act, or that the adjudicator was not appointed in accordance with the contractual provisions for appointment or did not satisfy the contractual requirement for a particular qualification. If the responding party does not complain about the appointment of the adjudicator he may say that the adjudicator has gone beyond the job that he was brought in to do.

These arguments may arise at the start of the adjudication process, part way through when for example the responding party seeks legal advice, or at the end when the successful claimant tries to enforce the decision. Even if it is not raised by the parties the adjudicator must address the question of jurisdiction. He will be concerned that if he proceeds without jurisdiction he may be wasting everyone's time and money. But he will also be concerned that his decision may be subjected to critical scrutiny by a court or arbitrator.

In the absence of any express provision in the Scheme or particular rules of the contract, it is necessary to return to first principles. Adjudication owes its authority to the terms of the contract, expressly set out or implied by the Scheme. If an adjudicator's decision is not made in accordance with the contractual system, it will not be enforceable by that contractual system, and will have no validity. The adjudicator's authority comes from the contract, and if he has no authority because he is operating outside the contract, his decision is worthless. The parties can however by their conduct rectify any gaps in the adjudicator's authority. Just as the parties can agree to refer to adjudication a matter which has nothing to do with a construction contract, they can also agree to ignore any potential lack of authority and be bound by the decision as if the appointment were entirely in accordance with the original contract.

The courts have had to deal with several cases involving jurisdiction arguments, and the decisions to date give valuable guidance as to how such matters can be resolved. It is clear that there are several potential ways to proceed.

The first case on jurisdiction was *The Project Consultancy Group* v. *The Trustees of the Gray Trust* (July 1999, Mr Justice Dyson) The plaintiff had made a claim for professional fees in connection with the conversion of a property into a nursing home. The claim had been referred to adjudication. The defendants had written to the adjudicator at the start of the proceedings in very clear terms to say that they protested the adjudicator's jurisdiction. They said that they would not recognise or comply with any decision that the adjudicator made. They also said that if the adjudication proceeded they reserved the right to participate, but such participation would be without prejudice to their argument that there was no jurisdiction.

Having made their point, they submitted to the adjudicator that the contract had been made before 1 May 1998, and that therefore the Act did not apply. They said that because of this there was no right to refer disputes to adjudication, and hence the adjudicator had no jurisdiction.

The adjudicator decided that he did have jurisdiction and went on to make a decision on the main issue that the plaintiff was entitled to be paid £64,975 plus VAT. The defendant did not pay, and so the plaintiff issued a claim in court and sought summary judgment under CPR Rule 24. The plaintiff argued that as there was an adjudicator's decision, that decision remained binding until determined by legal proceedings or agreement. Even though it was challenged, it should be enforced. Sir John Dyson rejected that:

'Accordingly, a decision purportedly made under section 108(3) in respect of a contract which is not a construction contract at all, or which is a construction contract entered into before Part II came into force, is not a decision within the meaning of the subsection, and is, therefore, not binding on the parties.'

He rejected arguments based on an analogy between the position of arbitrators at common law because the questions about adjudicators' jurisdiction were matters of statutory interpretation, not common law. He also dismissed the public policy argument that adjudicators should be given the authority to decide their own jurisdiction in order to avoid artificial arguments advanced to frustrate the adjudication process.

The judge did accept that the dissenting party could effectively submit the matter for decision to the adjudicator, and would then be bound by the decision. This had been established in a 1954 case on arbitration, *Westminster Chemicals & Produce Ltd* v. *Eicholz & Loeser*, and that principle would also apply to adjudication procedures, but the trustees had made their position entirely clear and had not agreed to submit to the adjudicator's jurisdiction on this or any other point.

Sir John Dyson was therefore prepared to consider the argument that the adjudicator had no jurisdiction because of the date of the contract. He was unable to do so in the context of an application for summary judgment, because the position was not sufficiently clear. On the facts he had to find that there was a real prospect that the defendant would establish that the contract had been concluded before 1 May 1998, and therefore refused the application.

The claimant fared rather better in *The Atlas Ceiling & Partition Co Ltd* v. *Crowngate Estates (Cheltenham) Ltd* (March 2000, Judge Thornton). There was a similar argument about jurisdiction based on the alleged date of the contract. The court action to enforce the adjudicator's decision was started on 31 January 2000. An immediate application was made for summary judgment, but once again there was sufficient doubt about the date of the contract to make summary judgment inappropriate. Instead the judge ordered that the hearing of the application would be the final trial of the issue, with cross examination of relevant witnesses. The trial took place on 18 February, less than three weeks from the date of issue, and rather earlier than most applications for summary judgment would have been heard. The claimant succeeded.

The *Project Consultancy* and *Atlas* cases dealt with the question of jurisdiction after the conclusion of the adjudication proceedings when the successful party was trying to enforce a decision in its favour. An alternative approach formed the subject of the decision in *Palmers Ltd* v. *ABB Power Construction Ltd* (August 1999, Judge Thornton). The facts of this dispute are described in Chapter 2 when considering the definition of 'construction operations', as ABB contended that the work done by Palmers did not fall within the statutory definition. Palmers had, on 15 July 1999, served a notice of adjudication and ABB had raised their argument five days later on 20 July.

Rather than proceed with the adjudication in the face of this argument, Palmers issued a claim in the Technology and Construction Court effectively seeking a declaration that the adjudicator had jurisdiction. The claim was issued on 26 July, and remarkably the case was heard on 30 July. Judgment was reserved, but nevertheless it was handed down on 5 August. Judge Thornton found that the contract was for construction operations, and therefore that the adjudicator had jurisdiction to deal with the dispute. He made it clear that he approved Palmer's decision to come to the court for an order:

> 'It is clearly appropriate for the court to intervene since it is only when it has declared that the relevant contract is a construction contract will an effective adjudication be possible. This is particularly so given that there is no statutory power given to an adjudicator, if appointed, to resolve disputes about his jurisdiction.'

The decision as to the jurisdiction of the adjudicator in the *Palmers* case was made by the court quickly enough to allow the adjudication to continue. There may be rather greater difficulties if the contract has an effective arbitration clause. In such a case the court might be prevented from making such a decision by an application to stay proceedings under section 9 of the Arbitration Act 1996. There is no reason why an arbitrator cannot deal with a question as to jurisdiction of an adjudicator very rapidly, and indeed some arbitrators practising in the construction law field have made very public statements about their willingness and ability to do so. The RICS maintains a panel of arbitrators who are prepared to commit to the production of an award within 48 hours of referral of the dispute. But before being able to make an award the arbitrator has to be appointed. Typical arbitration agreements provide for appointment of a person to be agreed by the parties, and in default of agreement within 14 days, a person nominated by a professional body such as the CIArb. If a jurisdiction point is taken at the start of an adjudication it may take more than two weeks to appoint an arbitrator. In the meantime the adjudicator will have to be making progress if he is to arrive at his decision within 28 days.

A third procedural route was considered in *Workplace Technologies v. E. Squared Ltd and Mr J.L. Riches* (February 2000, Judge Wilcox). Workplace, respondent in an adjudication, sought a declaration that the contract was formed prior to 1 May 1998, a declaration that the adjudicator had no jurisdiction, and an injunction to restrain and prevent E. Squared and the adjudicator from continuing with the reference. Once again the Technology and Construction Court was able to deal with the complete case quickly, and it was not necessary to consider an injunction pending the final decision. Judge Wilcox was not in any event persuaded that an injunction was appropriate in adjudication matters. He thought that in most adjudication cases it would be difficult to satisfy the balance of convenience test applied in injunction applications. After all, if a decision is given by an adjudicator without jurisdiction, the respondent can object to enforcement as in *Project Consultancy*.

Whilst the first cases involving jurisdiction points were relatively simple questions as to whether the contract was or was not within the ambit of the adjudication process, it has become clear that rather more complicated problems can arise. In *Homer Burgess Ltd* v. *Chirex (Annan) Ltd* (November 1999, Lord Macfadyan, Court of Session) the argument was whether part of the work could properly be classified as 'construction operations'. The pursuer in this Scottish case sought to enforce an adjudicator's decision regarding payment for works at a pharmaceutical processing and production facility. The defender said that a large proportion, but not all, of the works related to plant and was therefore within the exception created by section 105(2)(c)(ii) of the Act. The judge decided that the allocation of parts of a dispute into two categories, one of which was within the scope of the legislation and one of which was not, was a decision as to jurisdiction which would not be binding on a court. The adjudicator had decided that pipework linking various items of equipment at the site was not plant, and therefore not excluded from his jurisdiction. The court decided that he was wrong, and that therefore his decision with regard to that pipework was not to be enforced.

This seems to open up the possibility of an item-by-item challenge to an adjudicator's decision in the context of an application to enforce that decision. Lord Macfadyan also expressly reserved his position with regard to a challenge to a decision of fact by the adjudicator which would have the effect of placing the item inside or outside. In doing so he seems to raise the prospect of a more critical approach to adjudicator's decisions in Scotland than in England and Wales.

This divergence of approach was apparent in a decision given just three weeks after the *Homer Burgess* case in *Sherwood & Casson Ltd* v. *Mackenzie Engineering Ltd* (November 1999, Judge Thornton). The judge set out basic principles regarding the approach that the court should take in dealing with applications to enforce, which will be considered in more detail in Chapter 9. He relied in part on the decision of Mr Justice Dyson in *Bouygues UK Ltd* v. *Dahl-Jensen UK Ltd*, also considered in Chapter 9. One of the principles was stated thus:

> 'The adjudication is intended to be a speedy process in which mistakes will inevitably occur. Thus, the court should guard against characterising a mistaken answer to an issue, which is within an adjudicator's decision, as being as excess of jurisdiction.'

This philosophy would seem to be a major disincentive to the type of investigation which Lord Macfadyan contemplated in his reservation in *Homer Burgess*.

The *Sherwood & Casson* case was again an application to enforce an adjudicator's award, resisted on the ground that the adjudicator had acted in excess of his jurisdiction. It involved the argument that the dispute decided by the adjudicator was substantially the same as a dispute that had previously been referred to adjudication, and that therefore the adjudicator should have resigned as required by paragraph 9(2) of the Scheme. This question is different from the more usual problem of whether or not there was a construction contract in existence. As Judge Thornton explained:

> '... unlike the question of whether or not there is an underlying contract in existence, the adjudicator is given jurisdiction to determine whether or not the two disputes are substantially the same. The jurisdiction is analogous to that given to arbitrators for the first time by the Arbitration Act to determine their own jurisdiction.'

But that does not mean that the adjudicator's decision on the point is final:

> 'It makes no sense, as I see it, to impose on an adjudicator a mandatory requirement to resign if there is a substantial overlap between the dispute referred to him and one already decided by an earlier adjudication decision but then to make such an obligation unenforceable. This would be the effect of making an adjudicator's jurisdiction decision unchallengeable.'

Judge Thornton returned to the adjudicator's powers with regard to jurisdiction in *Fastrack Contractors Ltd* v. *Morrison Construction Ltd and Imreglio UK Ltd* (January 2000). His words provide a useful summary:

> 'If a party challenges the entire jurisdiction of the adjudicator, as Morrison does, it has four options. Firstly, it can agree to widen the jurisdiction of the adjudicator so as to refer the dispute as to the adjudicator's jurisdiction to the same adjudicator. If the referring party agrees to that course, and the appointed adjudicator accepts the reference to him of this second dispute, the jurisdiction of the adjudicator could then be resolved as part of the reference. The challenging party could, secondly, refer the dispute as to jurisdiction to a second adjudicator. This would not put a halt to the first adjudication, if that had already led to an appointment, since the adjudicator has a statutory duty, unless both parties agree otherwise, to decide the reference in a very short timescale. The challenging party could, thirdly, seek a declaration from the court that the proposed adjudication lacked jurisdiction. This option is of little utility unless the adjudicator has yet to be appointed or the parties agree to put the adjudication into abeyance pending the relatively speedy determination of the jurisdiction question by the court. The Technology and Construction Court can, for example, resolve questions of that kind within days of them being referred to it. Fourthly, the challenging party could reserve its position, participate in the adjudication and then challenge any attempt to enforce the adjudicator's decision on jurisdictional grounds. That is the course adopted by Morrison.
>
> The adjudicator can, of course, investigate any partial or entire jurisdictional challenge. He could, if he was satisfied that it was a good one, decline to adjudicate on the part of the reference he regarded as lacking jurisdiction. Alternatively, he could decide that the challenge was a bad one and proceed with the substance of the adjudication. However, unless the parties have vested the jurisdictional dispute in the hands of the adjudicator in addition to the underlying dispute, the adjudicator cannot determine his own jurisdiction and the challenging party may seek to avoid enforcement proceedings by showing that the sum claimed was decided upon without jurisdiction. The court would give appropriate weight to any findings of fact relevant to that jurisdictional challenge but would not be bound by them and would either have to bear out the challenge with evidence, or if that was not necessary, determine the challenge and either enforce or decline to enforce the whole part of the adjudicator's decision depending on the decision reached as to jurisdiction.'

Unless expressly permitted by the rules applicable to his appointment, the adjudicator must stay within his terms of reference, which will be established by the notice of adjudication. As discussed above, the contract rules governing the conduct of the adjudication or the Scheme may enable the adjudicator to expand the issues, but if the adjudicator strays outside the permitted bounds his decision on such wider matters will be of no effect. This was explained by Sir John Dyson in *Bouygues UK Ltd* v. *Dahl-Jensen UK Ltd* (November 1999), although the decision in that case was not in fact subject to that criticism:

> 'Where the adjudicator has gone outside his terms of reference, the court will not enforce his purported decision. This is not because it is unjust to enforce such a decision. It is because such a decision is of no effect in law.'

The TeCSA Rules are an example of a regime that does enable the adjudicator to rule on 'his own substantive jurisdiction'. If it can be established that he has been validly appointed, therefore, it will be difficult to challenge his decision on jurisdictional grounds.

Other sets of rules are less clear. Judge Thornton gave some helpful advice in *Christiani & Nielsen Ltd* v. *The Lowry Centre Development Co Ltd* (June 2000):

> 'It is prudent – indeed desirable – for an adjudicator faced with a jurisdictional challenge that is not a frivolous one, to investigate his own jurisdiction and to reach his own non-binding conclusion on that challenge.'

If the adjudicator concludes that he does not have jurisdiction, he will wish to advise the parties of his opinion. He may explain that if he is right his decision will be unenforceable, and before he continues to incur fees he will wish to know whether the parties wish him to do so. If only one of the parties wants him to proceed notwithstanding his concerns about jurisdiction it may be appropriate to advise that party that in those circumstances he may feel it appropriate to order that his costs be paid by that party.

If on the other hand the adjudicator concludes that he does have jurisdiction, he will proceed as normal. He may suggest that the question of jurisdiction be put to the court for an urgent decision, as in *Palmers Ltd* v. *ABB Power Construction Ltd*, or if there is an arbitration agreement he may suggest that an arbitrator be appointed to deal with the question quickly. He will not be able to require such action.

CHAPTER SIX
CONDUCT OF THE ADJUDICATION

6.1 Overriding duties of the adjudicator

The principal duty of the adjudicator under the Scheme is set out in paragraph 20:

'The adjudicator shall decide the matters in dispute'

This is what he has been appointed to do, and if he fails to do it he is failing the parties who have appointed him. The nature of this obligation will be considered further in Chapter 7. In addition to this fundamental duty, the Scheme imposes a timetable as is required by the Act, considered in detail later in this chapter, and at paragraph 12 the Scheme sets out some basic principles which are to govern the way in which the adjudicator conducts the process:

'12 The adjudicator shall
(a) act impartially in carrying out his duties and shall do so in accordance with any relevant terms of the contract and shall reach his decision in accordance with the applicable law in relation to the contract; and
(b) avoid incurring unnecessary expense.'

Section 108(2) of the Act requires all construction contracts to contain a clause imposing a duty on the adjudicator to act impartially (see Chapter 3), and so it is no surprise to find it in the Scheme. It is important that the adjudicator takes real pains to ensure that neither party becomes suspicious that he is acting other than impartially. Paragraph 4 of the Scheme bars any current employee of either party from being appointed, but a prior connection with one or other party does not disqualify the adjudicator from appointment. Before accepting appointment it is good practice for the adjudicator to ensure that both parties and, if relevant, the nominating body are aware of any connection that he may have had with either or both in the past. Paragraph 4 also requires the adjudicator to declare any interest, financial or otherwise, in any matter relating to the dispute.

The requirement that the adjudicator should act impartially is expressed in different terms under the Arbitration Act 1996, section 33 of which reads:

> '(1) The tribunal shall –
>   (a) act fairly and impartially as between the parties, giving each party a reasonable opportunity of putting his case and dealing with that of his opponent,'

but nevertheless the requirement of impartiality is the same. That requirement was examined in the 1999 case of *Laker Airways Inc* v. *FLS Aerospace Ltd*, which involved an application to remove an arbitrator on the basis that circumstances existed that gave rise to justifiable doubts as to his impartiality. Mr Justice Rix said that there were at least three principles at work. The first was that actual bias would always disqualify a person from sitting in judgment. The second was that no one could be a judge in his own cause or a cause in which he has a pecuniary or proprietary interest. This extended even to causes where the judge or other tribunal was very closely connected with a party as in the highly public case of *R* v. *Bow Street Magistrate, ex parte Pinochet (No 2)* (1999). There is no need to investigate whether or not there is a likelihood or even suspicion of bias in such cases.

The third principle cited by Mr Justice Rix was based on the judgment of Lord Goff in *R* v. *Gough* (1993). He quoted from that judgment:

> 'Accordingly, having ascertained the relevant circumstances, the court should ask itself whether, having regard to those circumstances, there was a real danger of bias on the part of the relevant member of the tribunal in question, in the sense that he might unfairly regard (or have unfairly regarded) with favour, or disfavour, the case of a party to the issue under consideration by him.'

The adjudicator should therefore not just satisfy himself that he is impartial, but also ensure that there is nothing in his conduct either of himself or of the adjudication that might give rise to a suspicion of bias or an allegation that he has not acted 'impartially in carrying out his duties'. The need to deal speedily with the dispute, and the very wide procedural discretion that he is given by the Scheme and the other sets of rules, make the possibility of misunderstanding the adjudicator's motives a serious risk, and great care needs to be taken. The decision of Judge Bowsher in *Discain Project Services Ltd v. Opecprime Developments Ltd* (August 2000), discussed in Chapter 9, suggests that the courts do expect adjudicators to avoid significant breaches of the principles of natural justice.

There is no procedure available to a party that considers that the adjudicator is failing to act impartially, comparable to section 24 of the Arbitration Act 1996 which enables an application to be made to the court to remove an arbitrator in such circumstances. It is possible that an application might be made to court for an injunction to prevent the biased adjudicator from proceeding with the adjudication. Alternatively the party might wait until the decision is made and then object to its enforcement. If the complaining party is in fact successful in the adjudication but considers that the decision in his favour was for an insufficient sum because of bias, he might make an application to court for a declaration that the decision is invalid so that a new adjudication can be commenced.

The second requirement of paragraph 12 of the Scheme is that the adjudicator should act in accordance with any relevant terms of the contract. Coupled with that is the requirement that he should reach his decision in accordance with the applicable law in relation to the contract. These requirements refer not only to procedure but also to the substance of his decision.

The adjudicator is later given wide discretion over procedural matters, but if the contract has a procedural requirement that does not conflict with that discretion, he must respect it. More significantly he must make a decision which is in accordance with the law. If acting under the Scheme, the adjudicator is not able to take a short cut to an answer by deciding what he considers to be fair or commercially reasonable. This is not the case under all institutional rules.

As with bias, it is not entirely clear what a party can do if he feels that the adjudicator is not acting in accordance with these rules. If he is able to take action before the decision is made it may be possible to obtain an injunction. Alternatively he may object to enforcement. As we will see though, when we deal with enforcement, a decision that is wrong may still be a valid decision and enforceable (as in *Bouygues UK Ltd v. Dahl-Jensen UK Ltd*, November 1999 and Court of Appeal July 2000), and there is no reason why that should not apply to incorrect law as well as to incorrect arithmetic. Failure to address the law and reliance on some other criterion may however render the decision unenforceable. It is unlikely that such a failure could be demonstrated if reasons are not given for the decision.

The adjudicator is required to avoid incurring unnecessary expense. This is one of two provisions limiting the right of the adjudicator to charge whatever he likes for his services. The other is found in paragraph 25 and will be considered in Chapter 8.

The various sets of institutional rules all impose a duty on the adjudicator to act impartially, as of course this is a requirement of the Act. The CIC Model Adjudication Procedure states clearly at paragraph 21:

> 'The Adjudicator shall determine the rights and obligations of the Parties in accordance with the law of the Contract.'

The JCT contracts do not state that the adjudicator has to make his decision in accordance with the contract, the applicable law or anything else, but it would be difficult to argue that such a requirement is not implied.

The TeCSA Rules go further than any standard procedures in giving the adjudicator flexibility to depart from strict law. Paragraph 15 of those rules provides:

> 'Wherever possible, the decision of the Adjudicator shall reflect the legal entitlements of the Parties. Where it appears to the Adjudicator impossible to reach a concluded view upon the legal entitlements of the Parties within the practical constraints of a rapid and economical adjudication process, his decision shall represent his fair and reasonable view, in the light of the facts and the law insofar as they have been ascertained by the Adjudicator, of how the disputed matter should lie unless and until resolved by litigation or arbitration.'

This paragraph will be a comfort to an adjudicator who is unable to carry out the exhaustive process expected of final arbitration or litigation in 28 days, but it will be impossible to tell when he has found it necessary to rely on it because the TeCSA Rules state at paragraph 27 that the decision shall not include reasons.

The ICE Adjudication Procedure does not state that the decision is to be in accordance with the law, but as with the JCT equivalent such a requirement can be implied. There is however a statement in paragraph 1.2 that:

> 'The object of adjudication is to reach a fair, rapid and inexpensive determination of a dispute arising under the Contract and this Procedure shall be interpreted accordingly.'

As has been explained in Chapter 3, there must be real doubt as to whether the ICE Procedure can ever apply, as the ICE contracts do not permit referral of disputes to adjudication 'at any time'. Moreover, as with the TeCSA Rules, it will be difficult to know whether or not the adjudicator has decided the dispute in accordance with the law, as reasons are not required.

6.2 The exercise of initiative by the adjudicator

Section 108(2) of the Act requires all construction contracts to 'enable the adjudicator to take the initiative in ascertaining the facts and the law'. This is carried through to the Scheme, which applies to any construction contract that has not met all the criteria of section 108. Paragraph 13 provides:

'The adjudicator may take the initiative in ascertaining the facts and the law necessary to determine the dispute, and shall decide on the procedure to be followed in the adjudication. In particular he may –
(a) request any party to the contract to supply him with such documents as he may reasonably require including, if he so directs, any written statement from any party to the contract supporting or supplementing the referral notice and any other documents given under paragraph 7(2) [the construction contract and any documents that the referring party intends to rely upon],
(b) decide the language or languages to be used in the adjudication and whether a translation of any document is to be provided and if so by whom,
(c) meet and question any of the parties to the contract and their representatives,
(d) subject to obtaining any necessary consent from a third party or parties, make such site visits and inspections as he considers appropriate, whether accompanied by the parties or not,
(e) subject to obtaining any necessary consent from a third party or parties, carry out any tests or experiments,
(f) obtain and consider such representations and submissions as he requires, and, provided he has notified the parties of his intention, appoint experts, assessors or legal advisers,
(g) give directions as to the timetable for the adjudication, any deadlines, or limits as to the length of written documents or oral representations to be complied with, and

(h) issue other directions relating to the conduct of the adjudication.

The adjudicator has an extremely wide discretion as to the procedures that he and the adjudication will follow. Some limit is imposed by the need to act impartially, in accordance with the contract terms, and without incurring unnecessary expense. Sub-paragraphs (d), (e) and (f) contain provisos to his ability to make visits, carry out tests and employ experts. Otherwise he has complete freedom.

Whilst the adjudicator has the power to do so, he is not obliged to take the initiative at all. He may invite the parties to agree a procedure and then follow it. Such an approach is unlikely to provide him with the information that he needs to enable him to make a decision within 28 days in any but the simplest of cases with unusually co-operative parties.

The 28 day period for the decision starts to run upon receipt of the notice of referral, but this does not mean that he cannot initiate procedures before that document is delivered. The notice of adjudication should give him a reasonable idea of the subject matter of the dispute, and he may feel able to give some preliminary directions as soon as he is appointed. At the very least he may wish to remind the claimant that the referral notice is due no later than seven days after the notice of adjudication, and should be accompanied by copies of the contract or relevant extracts from it and other documents that the claimant intends to rely upon. This is stated in the Scheme (paragraph 7), but it does no harm to repeat the requirement as a formal direction.

He may decide to require the respondent to submit a statement in response to the referral notice within a specified period, perhaps seven days. If he is considering an order of that sort, it is better to put the respondent on notice at the earliest possible moment. A respondent who has not experienced adjudication before, particularly if he is not normally involved in the construction industries, may not realise how fast adjudication has to proceed, and a firm statement by the adjudicator that he expects to see a fully documented submission within a week may come as a surprise.

It is unlikely that the adjudicator will feel able to go any further before he has seen at least the referral notice, but immediate preliminary directions can set the tone without actually providing for any definite further steps. For example, the adjudicator may state in his directions that he will call a meeting or hold a conference telephone call within two or three days of service of either the referral notice or the respondent's statement in order to give further procedural directions designed to enable him to arrive at his decision within 28 days. He may make it clear that while he will try to call the meeting at a time convenient for the parties, that may not be possible. Statements of that kind in preliminary directions make it easier to insist on a tight timetable later.

If the contract is one of the current JCT series or one of the associated subcontracts, the parties will have agreed to execute the JCT Adjudication Agreement, and the adjudicator may take this opportunity to require the parties to sign and return it to him.

Having received the referral notice, it may be appropriate to wait for the respondent's statement before doing anything further. The adjudicator is not obliged to do so, though, and he may feel that some time can be saved by anticipating questions that he expects to be relevant. For example, if the claimant is alleging that he has not been paid the price of variations to contract works, the adjudicator may consider that it will be important to establish when and how the variations were instructed. If this information is not provided in the referral notice, there is no reason why he should not ask the claimant to provide it straight away without waiting for the respondent to state his case on the point. He may also ask for further detail regarding the calculation of the sums claimed.

Similarly the referral notice and the documents that come with it may make it clear that there is a fundamental legal point in issue. There may be a dispute about the interpretation of the contract or the enforceability of a liquidated damages clause. If the adjudicator is aware of a legal argument that will have to be addressed by the parties he may require the claimant to prepare submissions on the point and also require the respondent to deal with it in the response that is currently being prepared.

The adjudicator may, within reason, ask to see any document. The document might be directly relevant to the dispute, but there is no restriction. If the dispute involves an allegation by a main contractor that a subcontractor has caused delay to the main contract works, the adjudicator may decide that he wishes to investigate other aspects of progress on the site. He may ask to see correspondence between the main contractor and other subcontractors and between the main contractor and the architect. He does not need to wait for the subcontractor to ask to see the documents.

These are examples of the exercise by the adjudicator of the power suggested in paragraph 13(a) of the Scheme. In traditional arbitration, such behaviour would have been considered unconventional, but an adjudicator should not be deterred if he thinks it appropriate.

The list of powers set out in paragraph 13 of the Scheme is not designed to be a comprehensive list of things that the adjudicator can do, but merely illustrations of the sort of steps that he might take, with a few cautionary provisos.

Language is unlikely to be a major concern in adjudication proceedings, but it must be remembered that the parties may not be English speaking. With the increasing internationalisation of the construction market, it is at least possible that the dispute may be between a contractor and a subcontractor neither of whom is based in the UK. Even if the parties are happy that the proceedings are conducted in English, there may be relevant documents in different languages, requiring translation. The adjudicator has authority to order a party to provide the translation, or he can decide to have a document translated by an independent agency. That authority is implicit in the general provision of paragraph 13, but it is also expressly provided in subparagraph 13(b).

The adjudication may be conducted on the basis of documents only, or the adjudicator may require to meet the parties. These meetings may take the form of formal hearings, as in conventional legal proceedings, or they may be more in the nature of an inquiry with the adjudicator taking the lead role. If the adjudicator decides to take a less formal approach he will have to take real care to keep control. It may be very helpful to allow a dialogue to develop between the parties in response to a specific question from the adjudicator, and the discussion may reveal more than the initial response to the adjudicator's question, but it is easy for the discussion to degenerate into a shouting match.

The adjudicator may decide to have a series of meetings dealing with different issues or groups of issues. He may hold the meeting(s) wherever he feels it most appropriate, which may be in his own office, on site or in the offices of one of the parties. The meeting(s) might be held at any time of the day or night. The adjudicator may hold meetings with only one of the parties if he considers it appropriate. This might be necessary if it is not possible to make arrangements for a meeting at a time when both can attend, but the adjudicator should be careful to comply with the obligation in paragraph 17, considered below.

The adjudicator may decide that it would be helpful to visit the site either to familiarise himself with the layout and features or to inspect some item of work. Subparagraph 13(d) reminds the adjudicator that he may be trespassing as neither of the parties to the dispute may be able to authorise him to have access, and so consent may have to be obtained. A visit to the site may be invaluable if work is continuing, and if the dispute is about quality of work or measure an inspection may be essential. There is no reason why the adjudicator should not carry out his visit or inspection without the parties being present, so long as he remembers paragraph 17.

Similarly a test or experiment may be needed in a technical dispute and there is no reason why the adjudicator, if appropriately experienced, should not carry it out himself. Once again subparagraph 13(e) reminds the adjudicator that he may need consent from someone other than the parties, and he will have to comply with paragraph 17.

The adjudicator cannot delegate the decision-making process to someone else, but he can take advice from practitioners in fields other than his own. Subparagraph 13(f) states that he can appoint 'experts, assessors or legal advisers', and if the subject matter of the dispute involves an area in which he is not experienced he should seriously consider exercising this power. It is not always necessary to do so. Adjudicators listed on panels maintained by the adjudicator nominating bodies have had to satisfy the relevant institution that they understand the legal principles of construction contracts and the adjudication process, and even though not legally qualified they should be able to deal with the legal issues arising in most construction contract disputes. But if there is a particularly difficult problem it would clearly be more satisfactory for the non-lawyer adjudicator to take legal advice. A lawyer confronted with a technical issue may be able to decide on the basis of submissions from experts retained by the parties, but again it may be more satisfactory and substantially more economical to consult an independent consultant. If there are substantial measurement and valuation issues a quantity surveyor assessor may be essential. If the adjudicator is going to consult an independent expert he must decide to do so very early on in the process, as he may be restricted to the 28 day limit in which to instruct the adviser, receive the advice and incorporate it in his decision.

Subparagraph 12(f) imposes a restriction on the adjudicator's powers to instruct an independent expert – he must advise the parties of his intention to do so before he appoints. This is not just to warn the parties to expect an additional charge. The adjudicator should advise the parties not just of his intention to appoint an expert, but also who the expert is and the nature of the assistance sought. This enables the parties to make representation as to the suitability of the expert. One of them may strongly object to the person chosen because of some previous connection. The parties may also feel that the adjudicator is asking the wrong question, and giving them an opportunity to say so at the start may save valuable time and cost.

The general discretion of the adjudicator to decide procedure clearly includes the power to decide the timetable, but subparagraph 13(g) goes further and suggests that the adjudicator may wish to impose limits on the time to be taken in oral representations and the length of written documents. The adjudicator will have in mind the need to give the maximum opportunity to both sides to present their case fully but will also be aware of the need to give each side a realistic chance of responding to the other's case. Those two objectives have to be achieved within a very tight timescale – 28 days unless the claimant agrees to extend it.

In the *Macob* case, discussed in more detail below, Sit John Dyson commented that:

> 'It is clear that Parliament intended that the adjudication should be conducted in a manner which those familiar with the grinding detail of the traditional approach to the resolution of construction disputes apparently find difficult to accept.'

It is certainly true that parties who have not experienced the process before, but perhaps have occasionally been involved in arbitration or litigation, may find it difficult to accept the timetable that the adjudicator will direct, and it may be necessary for the adjudicator to flex his muscles. It is not unusual for the adjudicator to have to point out that evenings, weekends and bank holidays are available for the preparation of statements and submissions as well as normal working hours.

Subparagraph 13(h) is a reminder that the powers set out in the previous subparagraphs are not intended to be a comprehensive statement of what the adjudicator can do. He has complete discretion, so long as he acts impartially, in accordance with any relevant terms of the contract and avoids incurring unnecessary expense. He might, for example, speak to an independent person who he thinks may be able to assist in providing relevant evidence. In a typical dispute between a subcontractor and the main contractor, the clerk of works or another subcontractor may be very helpful in establishing what was really going on. If the adjudicator is going to speak to third parties in this way it would be advisable to make a careful record of what he is told and copy it to both parties, but there is no absolute requirement on him to do so.

Reference has been made in several paragraphs above to the provision in paragraph 17 of the Scheme which reads:

'17. The adjudicator shall consider any relevant information submitted to him by any of the parties to the dispute and shall make available to them any information to be taken into account in reaching his decision.'

Some adjudicators are tempted to state that if a requested statement is not delivered to them by a specific time or date it will not be considered. It is not open to the adjudicator to refuse to consider any information presented to him providing that it is relevant, but if information is supplied late he may attach little weight to it (see paragraph 15, discussed below). Any information given to him that he is going to take into account must be made available to all the parties. He will normally direct that all communications sent to him should simultaneously be sent to the other parties, but this will not cover the product of his own investigations, tests or experiments, or the advice given to him by independent consultants retained as experts. All such information must be disclosed to the parties. The paragraph does not say when this information has to be made available, but there is little point in revealing it when or after the decision is made. If at all practicable the information should be passed to the parties in time for them to comment on it before the adjudicator makes his decision.

Unlike some other sets of adjudication rules, discussed below, which state that the adjudicator may make use of his own knowledge and expertise, the Scheme does not do so. There is no reason why the adjudicator should not draw on his own resources in this way, and indeed he may well have been selected for appointment on the assumption that he will do, but if he is drawing on his own specialist knowledge it would be helpful to the parties to make this known to them.

The equivalent provision of the JCT series of contracts is different from the Scheme in several respects. The adjudicator once again has absolute discretion in setting the procedure and he is able to take the initiative in ascertaining the facts and the law. In addition it is stated that he may use his own knowledge and/or experience. He may require the parties to carry out tests or open up the work rather than carry out the tests himself and he may visit relevant workshops as well as the site. He may obtain information from an employee of a party instead of the party itself so long as prior notice is given to the party. Before consulting an independent adviser, the adjudicator must give prior notice to the parties as under the Scheme, but must also give them a statement or estimate of the cost involved.

The TeCSA Rules contain further differences. They give the

adjudicator the power, and indeed the duty, to establish the procedure and timetable for the adjudication, and as with the Scheme they set out a number of examples of how his discretion might be exercised.

One suggestion is that the adjudicator may:

'Require any party to produce a bundle of key documents, whether helpful or otherwise to that Party's case, and to draw such inference as may seem proper from any imbalance in such bundle as may become apparent.'

Adjudication as a process is far removed from court procedures and is by its nature rather more confrontational than litigation in the post-Woolf era, but this suggested direction seems to owe much to the 'cards on the table' approach of the Civil Procedure Rules. Technically there is no reason why a similar order could not be made in a Scheme adjudication, or indeed an adjudication under any other set of rules.

As with the Scheme, the TeCSA Rules suggest that the adjudicator may require the delivery of documents, but unlike the Scheme these rules exclude from this 'documents that would be privileged from production to a court'. A letter written by the respondent, for example, making a 'without prejudice' offer to settle, is not something that the adjudicator can request. Subparagraph 13(a) did not however suggest an unlimited right to require documents, but was restricted to 'such documents as he may reasonably require'. Arguably it would be unreasonable to require documents that are privileged from production.

The TeCSA Rules expressly allow the adjudicator to make use of his own specialist knowledge.

When acting under the TeCSA Rules, the adjudicator may only retain an independent consultant if at least one of the parties so requests or consents, but there is no obligation to reveal the substance of the advice to the parties.

The CIC Rules authorise the adjudicator to obtain independent technical or legal advice, provided that he has given notice to the parties of his intention to do so. Under these rules he is obliged to provide to the parties copies of written advice received.

6.3 Failure to comply

Paragraphs 14 and 15 of the Scheme deal with compliance. Paragraph 14 states the rule:

> '14. The parties shall comply with any request or direction of the adjudicator in relation to the adjudication.'

Paragraph 14 is useful if confirming the authority of the adjudicator over the procedure of the adjudication. It was argued in *Macob Civil Engineering Ltd* v. *Morrison Construction Ltd* (February 1999, Sir John Dyson) that this authority was subject to the overriding obligation to comply with the rules of natural justice. In the first reported case dealing with the enforceability of an adjudicator's decision, one of the defendant's arguments was that the decision was invalid and unenforceable because the adjudicator was guilty of procedural error in conducting the adjudication in breach of those rules.

The adjudicator had been dealing with a contested notice of intention to withhold a payment. The parties had each argued that there were agreements in place about when money was to become due. He was unable to decide what had been agreed, and therefore concluded that the parties had failed to provide an adequate mechanism for determining dates when payments were to become due and the final date for payment, as required by section 110(1) of the Act. That led him to decide that the provisions of the Scheme applied. Morrison had not given proper notice of intention to withhold under the Scheme. Morrison argued that the adjudicator should have asked the parties to make representation to him about whether or not there was an adequate mechanism for payment in the contract, and his failure to do so was a breach of the rules of natural justice.

Morrison also complained of a breach of natural justice in that the adjudicator had invoked section 42 of the Arbitration Act 1996, expressing his decision as a peremptory order, without giving the parties an opportunity to make representations on that point either.

Sir John Dyson did not actually decide whether there had been a breach of the rules of natural justice by the adjudicator. He said that he formed a strong provisional view that the challenge to the decision on that basis was 'hopeless', but in his view he did not need to decide the point. Even if the adjudicator had made a procedural error which invalidated the decision, the decision would still be a decision and would therefore be enforceable.

Sir John Dyson did accept that the adjudicator was required to comply with paragraph 12(a) of the Scheme:

> '12. The adjudicator shall –
> act impartially in carrying out his duties and shall do so in accordance with any relevant terms of the contract and shall

reach his decision in accordance with the applicable law in relation to the contract;'

But otherwise, as he said, 'The adjudicator is given a fairly free hand.' The decision in *Macob* made it clear that decisions of adjudicators are unlikely to be rendered unenforceable by procedural considerations. The more recent decision in *Discain Project Services Ltd* v. *Opecprime Developments Ltd* (August 2000 – see Chapter 9) suggests that the adjudicator must avoid significant breaches of the rules of natural justice, but nevertheless it is therefore prudent for parties to comply with the adjudicator's directions.

In fact there is little that the adjudicator can do to compel such compliance. Paragraph 15 of the Scheme sets out some suggestions as to how he can deal with a failure by either party to do what he is told:

'15. If, without showing sufficient cause, a party fails to comply with any request, direction or timetable of the adjudicator made in accordance with his powers, fails to produce any document or written statement requested by the adjudicator, or in any other way fails to comply with a requirement under these provisions relating to an adjudication, the adjudicator may –
   (a) continue the adjudication in the absence of that party or of the document or written statement requested,
   (b) draw such inferences from that failure to comply as circumstances may, in the adjudicator's opinion, be justified, and
make a decision on the basis of the information before him attaching such weight as he thinks fit to any evidence submitted to him outside any period he may have requested or directed.'

If the adjudicator has directed that the parties attend a meeting, for directions, to discuss or answer questions about any matter in dispute, or for a formal hearing, and one party fails to attend, paragraph 15(a) makes it clear that the adjudicator can carry on regardless. This does not mean that he should do so. The adjudicator must make up his mind in the light of the circumstances at the time. If there is little time left for the decision to be made it may be necessary to carry on in a party's absence. If the meeting is at the start of the process the adjudicator can afford to be a little more forgiving. If time permits the adjudicator may wish to make some enquiry to find out why the party has not appeared, but that may be quite obvious. For example, the non-appearing party may have made it clear that he disputes the adjudicator's jurisdiction and will not be taking any part in the process.

Similar questions arise when a party fails to produce a statement (such as a response to the referral notice) or a document that the adjudicator has asked to see. If the statement or document does not turn up on time the adjudicator can press on, but he may prefer to make some enquiry to find out why the document has not been produced and give the defaulting party a second chance.

The second and third subparagraphs of paragraph 15 are helpful in suggesting how the adjudicator can proceed in the face of non-compliance without completely debarring a party or its statement or submission.

The first suggestion is that the adjudicator may draw inferences from the failure. If a party fails to provide details of a particular head of claim, or copies of documents that support it, the adjudicator may be justified in concluding that there are none, and proceed accordingly. If an aspect of the case depends on the evidence of a specific person who fails to attend a meeting as requested by the adjudicator, the adjudicator may be justified in concluding that the relevant head of claim is unsupportable.

Subparagraph (c) makes it clear that the adjudicator can proceed on the basis of the information that is available to him. He has the power to take the initiative in ascertaining the facts and the law, as stated in paragraph 13, but he does not have to carry out a comprehensive enquiry. He has to do the best he can with the material provided to him, and if one or other of the parties fails to provide him with its best case it only has itself to blame.

Given the extremely short timetable for the complete process it is inevitable that a party will occasionally fail to meet deadlines ordered by the adjudicator. It is also inevitable that the other party will try to take advantage of this failure and ask the adjudicator to strike out part or all of the case. This may be justified, and subparagraph 15(a) states that this is one option available to the adjudicator. But subparagraph 15(c) offers an alternative. He can accept the late submission, 'attaching such weight as he thinks fit' to it. The important question is whether or not the other party has time to consider the late information, and whether that party is put at a serious disadvantage because of the delay. If there is insufficient time to probe the information and present a response, the adjudicator must consider whether it would have made any significant difference if the information had been provided on time. If it seems that it may have made such a difference the evidential value of the late material is clearly lessened. It does the process no harm for the adjudicator to remind the parties of the suggestions of this subparagraph when procedural delay is threatened. A party who says that he cannot provide a statement within 48 hours as directed often finds new energy when he is told that delay will lead to less weight being attached to it.

The adjudicator must however remain flexible. Adjudication is not arbitration, nor is it post-Woolf litigation with commitment to procedural timetables. The only timetable that really matters is the overall period of 28 days, possibly extended. If there are good reasons for a failure to comply with directions the adjudicator would be wrong to punish such failure by proceeding in the absence of a party, by drawing negative inferences or discounting the value of evidence. He is still required by paragraph 17 to consider any document placed before him. The adjudicator must use his powers creatively to assist him in obtaining from the parties the information that is necessary to enable him to make a decision. How he does this is a matter for his discretion. It is unlikely that his decision would be rendered unenforceable because he drew an inference which does not seem justified to someone else.

The JCT series of contracts and subcontracts includes provision that any failure of either party to comply with the adjudicator's directions 'shall not invalidate the decision of the Adjudicator'. This does not really give the adjudicator any real assistance when facing an unco-operative party, but there is nothing to suggest that the adjudicator cannot act in the same way as he would under the Scheme. The other published sets of adjudication rules do not deal with non-compliance at all, and again the adjudicator should consider himself free to act as he thinks appropriate, subject to the requirement of impartiality.

6.4   Representation of the parties

Paragraph 16 of the Scheme states:

'16–(1) Subject to any agreement of the parties to the contrary, and to the terms of paragraph (2) below, any party to the dispute may be assisted by, or represented by, such advisers or representatives (whether legally qualified or not) as he considers appropriate.

(2) Where the adjudicator is considering oral evidence or representations, a party to the dispute may not be represented

by more than one person, unless the adjudicator gives directions to the contrary.'

There is therefore no restriction on who may present a case, written or oral, in adjudication, save for the limit on numbers in sub-paragraph (2).

In the period of discussion before and after the passing of the Act, and before the process of adjudication as we now know it started to operate, it was assumed by some that adjudication would remove the need for lawyers in the process of dispute resolution in the construction industry. As many pre-Act disputes were resolved in arbitration rather than litigation, it might be argued that there was never a need for lawyers in most formal disputes, and indeed many arbitrations have proceeded without any representation by lawyers, claims consultants or anyone external to the parties themselves. Nevertheless many parties preferred to retain lawyers to manage the preparation of cases and the presentation of them before arbitrators as they would in court.

Lawyers can help or hinder the adjudication process, and the same is true of other consultants. A party to the dispute may not fully understand the issues, legal and factual, that lie behind his position. He may know that he feels aggrieved commercially and he may believe that he has been underpaid or blamed for something that is not his responsibility, but in order to persuade an adjudicator to find in his favour he must be able to express himself in contractual terms. The sense of grievance will not get him very far unless he can demonstrate a legal entitlement. Moreover the case must be made to seem clear and simple. The adjudicator has a very short time in which to make a decision, and whilst he may be prepared to carry out an exhaustive enquiry he will be more likely to find in favour of a party who has made exhaustion unnecessary.

There are cases where this analysis and clear presentation is not complicated. In such cases the party is probably best advised to represent himself. If he would like a second opinion before committing himself formally a brief discussion with a lawyer or claims consultant may be a good investment, but the expense of a legal representative throughout the adjudication will probably not be justified, particularly as that expense is likely to be irrecoverable (see Chapter 8).

On the other hand some cases really do require external assistance. The referral notice has to be prepared with real care, including enough detail to be convincing but remaining clear and easy to follow. The response will have to be prepared with remarkable speed, and clarity will be even more difficult to achieve. The adjudicator may need help in concentrating on the important issues, and may also have some unconventional ideas about procedure that are less than helpful to the presentation of the case. A lawyer or other consultant with experience of construction disputes and in particular of the adjudication process may then be invaluable, but only if he is able to commit himself to the demands of the adjudication timetable.

None of the published standard rules of adjudication restrict the right of a party to employ a representative of his choice, and none even goes so far as the Scheme in expressly suggesting a limit of representation in hearings. The TeCSA Rules imply a similar limit though by stating that:

'21 The Adjudicator may not...
iii) Refuse any party the right at any hearing or meeting to be represented by any representative of the Party's choosing who is present.'

The implicit suggestion is that only one representative can represent the party at any one time.

When retaining lawyers or others to advise or represent a party in adjudication, it must be remembered that the professional fees incurred are unlikely to be recovered. Costs are discussed in detail in Chapter 8.

6.5 *Confidentiality*

Parties to an adjudication under the Scheme can take steps to protect confidentiality. Paragraph 18 of the Scheme provides:

'The adjudicator and any party to the dispute shall not disclose to any other person any information or document provided to him in connection with the adjudication which the party supplying it has indicated is to be treated as confidential, except to the extent that it is necessary for the purposes of, or in connection with the adjudication.'

If a party is disclosing information that it wishes to keep private, it must make that clear, preferably at the time that it is supplied. If that is done, there is a requirement that the adjudicator and the other party or parties should comply with the request and keep it confidential. That requirement is a contractual matter, as is the adjudication process itself, and if the confidentiality is not respected the remedy is a claim for damages for breach of contract, or in serious cases where prevention of disclosure is required, an injunction might be obtained.

There are clearly many documents and types of information that an adjudicator may require to see which a party may feel are highly sensitive. The profit margin built into a tender price is likely to be highly relevant to issues regarding valuation of variations or loss and expense claims. The contractor or subcontractor may be very reluctant to disclose all the relevant documents. In conventional litigation or arbitration the matter may be rather less sensitive because the time for disclosure is several months or years after the event, but adjudication is likely to be more immediate. The adjudicator has the power to require production of this sort of information. Commercial sensitivity is not sufficient excuse for refusal to comply with such a requirement.

Moreover it is not possible to produce a document to the adjudicator stating that it is for his eyes only. Paragraph 17 states that the adjudicator shall make available to the parties any information to be taken into account in reaching his decision. This applies whether or not the adjudicator is required to give reasons for the decision. Any document or other information disclosed to the adjudicator will therefore find its way to the other side unless it is completely irrelevant.

When producing any document in adjudication proceedings, including any statement such as the notice of referral or written evidence, consideration should be given to making an express statement that the document is confidential. When the adjudication is under the Scheme, there is then a contractual reason for expecting the confidentiality to be respected.

If the adjudication is proceeding under the CIC Rules, and the parties and the adjudicator have signed the CIC Adjudication Agreement, there is again some contractual protection for confidentiality, in that clause 4 of the agreement provides:

'The Adjudicator and the Parties shall keep the adjudication confidential, except so far as is necessary to enable a Party to implement or enforce the Adjudicator's decision.'

This appears to relate just to the adjudication proceedings themselves, and not to information obtained during the course of the proceedings, but a court would be unlikely to have much sympathy for such sophistry.

The TeCSA Rules include a more comprehensive confidentiality provision:

> '30. The Adjudication and all matters arising in the course thereof are and will be kept confidential by the Parties except insofar as necessary to implement or enforce any decision of the Adjudicator or as may be required for the purpose of any subsequent proceedings.'

This clause however differs from the others in that it does not require the adjudicator to keep matters confidential.

The Adjudicator's Appointment under GC/Wks/1 requires the adjudicator to comply with the Official Secrets Act 1989 and, where appropriate, section 11 of the Atomic Energy Act 1946. Any information concerning the contract is stated to be confidential. Confidentiality between the parties is not dealt with by the adjudication provisions, being covered elsewhere in the contract.

The JCT series of contracts and subcontracts do not make any express provision regarding confidentiality.

6.6 Timetable for decision

The timetable for the decision in Scheme adjudications is established by paragraph 19(1), which gives effect to the relevant parts of section 108(2) of the Act:

> '19- The adjudicator shall reach his decision not later than –
> (a) twenty eight days after the date of the referral notice mentioned in paragraph 7(1);
> (b) forty two days after the date of the referral notice if the referring party so consents; or
> (c) such period exceeding twenty eight days after the referral notice as the parties to the dispute may, after the giving of that notice, agree.'

The 28 day period does not run from the date of the notice of adjudication, or the date when the adjudicator is appointed, but from the date that the referral notice is received by the adjudicator. The referral notice should have been delivered within seven days from the notice of adjudication, but if it is delivered late, the start of the 28 day period is put back.

The timetable is one of the few aspects of the adjudication proceedings that is not within the control of the adjudicator. He cannot in any circumstances give himself an extension of time. If the adjudicator needs more than 28 days to reach his decision, he must first ask the claimant for an extension. The claimant can grant an extension of 14 days, making 42 days in total.

If that is not sufficient, and the adjudicator needs more time again, all the parties to the adjudication must agree how much time is to be given to him. It is said both in the Scheme and in the Act that the agreement must be made after the delivery of the referral notice. This prevents any attempt to prolong the adjudication proceedings by including an extension of time agreement in, for example, a main contractor's standard form of subcontract.

Section 116 of the Act excludes Christmas Day, Good Friday and bank holidays from the reckoning of periods of time under the Act, but there is no corresponding provision in the Scheme. Hence the periods set out in paragraph 19 cannot be read as excluding any calendar days at all.

The claimant will of course want the adjudicator to reach his decision as soon as possible. In practice though he is unlikely to be difficult about the adjudicator's request for the initial 14 day extension. He will normally be anxious not to upset the adjudicator for obvious reasons, and the adjudicator is likely to suggest that the extra time is needed so that he can give proper consideration to the points being made by the claimant.

Rather different considerations apply to the further extension, requiring agreement by both (or all if more than two) parties. Unless the dispute is of very great complexity, the most likely reason for the adjudicator requesting more than 42 days is that one party is in difficulty in producing information requested by the adjudicator. The other party may wish to keep up the pressure by refusing consent to an extension that might assist the first party out of its difficulty. If the adjudicator believes that it is right that more time should be given he may well have to use all his powers of charm and persuasion.

As we have seen above, the adjudicator has complete discretion, subject to the requirement of impartiality, in dealing with the timetable for the adjudication within the overall time requirements of paragraph 19(1). If he does not produce his decision within the 28 days (or the agreed extended time), paragraph 19(2) applies:

> '19–(2) Where the adjudicator fails, for any reason, to reach his decision in accordance with sub-paragraph (1) –

(a) any of the parties to the dispute may serve a fresh notice under paragraph 1 and shall request an adjudicator to act in accordance with paragraphs 2 to 7; and

(b) if requested by the new adjudicator and in so far as is reasonably practicable, the parties shall supply him with copies of all documents which they had made available to the previous adjudicator.'

If the adjudicator resigns, paragraph 9 of the Scheme applies and the referring party can start a new adjudication. If however the adjudication comes to a halt because the adjudicator is indisposed or for some reason refuses to produce a decision, but does not resign, paragraph 19(2) applies and either party can start afresh.

This gives scope for considerable confusion. It is all very well for a new adjudicator to be appointed when the original adjudicator has resigned, but if he has not resigned he may well consider that he is still supposed to be acting. He may still produce a decision, albeit late, and expect to be paid. It would be possible for the parties to revoke the first appointment under paragraph 11, but one party may prefer the original adjudicator to the new one.

If the first adjudicator has not resigned and his appointment has not been revoked he remains in office as the adjudicator, notwithstanding that his decision has not been produced on time. The appointment of the new adjudicator does not automatically revoke the first appointment. If the first adjudicator then produces a decision it will still be a 'decision', even though late. The approach of Sir John Dyson in *Macob Civil Engineering Ltd* v. *Morrison Construction Ltd*, discussed earlier in this chapter, suggests that procedural irregularity does not prevent a decision from being a decision and therefore enforceable:

'If his decision on the issue referred to him is wrong, or because in reaching his decision he made a procedural error which invalidates the decision, it is still a decision on the issue.'

As we shall see in Chapter 7, the adjudicator's decision is binding until the dispute is finally determined by legal proceedings, by arbitration or by agreement between the parties. The second adjudicator, who no doubt will find out about the first adjudicator's decision during the course of the second adjudication, will have little choice but to conclude that the parties are bound by the first decision, even if he disagrees with it.

The practical consequences of being a few days late in producing a decision are therefore unlikely to be severe, unless both parties are sufficiently dissatisfied with the adjudicator's performance that they wish to revoke his appointment under paragraph 11. Failure to comply with the timetable provision of paragraph 19 would be a default on the part of the adjudicator and revocation in those circumstances would mean that the parties would not be liable to pay the adjudicator's fees, as provided by paragraph 11(2).

The last provision of the Scheme dealing with the timetable is paragraph 19(3):

'(3) As soon as possible after he has reached a decision, the adjudicator shall deliver a copy of that decision to each of the parties to the contract.'

If he is able to arrive at his decision before the expiry of the time limit, the adjudicator cannot hold on to it until the 28 days is up. The main reason why the adjudicator may wish to delay delivery of the decision is to ensure that his fees are paid, as has long been the custom in arbitration. Typically an arbitrator will write to the parties saying that his award has been published and is available for collection on payment of the outstanding fees. He exercises a lien over the award. This is clearly not possible in adjudication under the Scheme.

Most of the other published sets of rules governing adjudication do not differ in substance from the Scheme on the point of timetable for reaching a decision, as of course the timetable was established by the Act. Clause 59(5) of GC/Wks/1 imposes a 'not before' date as well as the normal time limit:

'The adjudicator shall notify his decision to the PM, the QS, the Employer and the Contractor not earlier than 10 and not later than 28 Days from receipt of the notice of referral...'

The CEDR rules also differ slightly from the norm in that clause 9 provides that:

'The Adjudicator shall reach a decision as soon as practicable, the objective being to have a decision within 14 days of the date of referral.'

Having set out this admirable objective, CEDR then imposes the same time requirements as the Scheme.

Care is needed when dealing with the JCT contracts. The 1998 versions of the series set out the timetable requirements in accordance with the Act, but the amendments to the previous editions published as supplements prior to the publication of the 1998 editions did not comply. They required delivery of the referral to the adjudicator within seven days of the notice of adjudication or the execution of the JCT Adjudicator's Agreement, whichever was the later.

There are more differences between the various sets of rules in dealing with what happens if the adjudicator fails to meet the deadline for producing his decision.

The JCT system does not say what will happen if the decision is not produced on time, but under the JCT Adjudication Agreement (considered in detail below), the parties can terminate the agreement (and therefore the appointment of the adjudicator) at any time. This would require agreement of the parties, which as suggested above may not always be easy to achieve. If the termination is because of a failure to give the decision within the time-scale the adjudicator does not recover his fees and expenses.

Under the ICE Adjudication Procedure, either party may refer the dispute to a replacement adjudicator if the first has failed to produce a decision in time, providing it gives seven days' notice of intention to do so. Unlike the Scheme, the ICE Procedure deals with what happens if the first adjudicator carries on regardless. If he produces his decision late, but before the dispute has been referred to a new adjudicator, the decision is effective and he is paid. Once the dispute has been referred to a replacement, however, the decision is of no effect. In those circumstances the first adjudicator is not entitled to be paid fees and expenses, save for the cost of independent technical or legal advice properly obtained.

GC/Wks/1 expressly states that the adjudicator's decision is valid even if late. Under the Adjudicator's Appointment for use with this contract, the adjudicator agrees to comply with the time limits set out in condition 59, but apart from the question of validity, nothing is said about what happens if he fails to do so.

No provision is made under the TeCSA Rules for failure by the adjudicator to produce his decision on time, but the chairman of TeCSA has power to replace any adjudicator nominated by him if and when it appears necessary to him to do so. Either or both parties can complain to him about failure of the adjudicator in several respects, including failure to proceed with the adjudication with necessary despatch.

The CIC Rules are similar to the ICE Procedure in that they state that if the adjudicator fails to reach his decision within the time permitted, either party can request the appointment of a replacement adjudicator. If the original adjudicator produces his decision before referral to the replacement, it is effective. If he fails to make an effective decision, he is not entitled to be paid his fees and expenses except the cost of legal or technical advice which has been received by the parties.

6.7 Standard forms of appointment

Several standard contracts and sets of adjudication rules referred to or contained within those contracts provide for the execution by the parties and by the adjudicator of a standard form of appointment.

6.7.1 The JCT contracts

The JCT series of contracts and associated subcontracts provide that no adjudicator shall be agreed or nominated who will not execute the JCT Adjudication Agreement, and when the adjudicator has been agreed or nominated, the parties and the adjudicator are to execute the agreement. There are only five clauses of the agreement:

(1) The parties appoint the adjudicator and the adjudicator accepts the appointment
(2) The adjudicator agrees to observe the adjudication provisions set out in the contract between the parties
(3) The parties agree to be jointly and severally liable for the fees and expenses of the adjudicator
(4) The adjudicator agrees to tell the parties if he becomes ill or otherwise unavailable to complete the adjudication
(5) The appointment is terminable on notice by the parties to the adjudicator. On such termination the parties will pay the adjudicator's fee, unless the termination is because the adjudicator has failed to give his decision within the time-scale or at all.

These provisions are of no real surprise. The agreement is of more value in setting out a number of matters in the recitals and the schedule. The parties are stated with full names and addresses at the start of the document, together with the details of the adjudicator. The recitals record the details of the contract under which the dispute has arisen, including the form of contract and any amendments. The schedule states how the adjudicator's fee is to be calculated, either on the basis of a fixed lump sum or an hourly rate. Provision of this sort of information at the start can avoid considerable debate later. The JCT contracts do however anticipate that the agreement may not always be signed. They state that a failure of a party to enter into the adjudication agreement shall not invalidate the adjudicator's decision. The contractual relationship between the parties and the adjudicator is not fundamentally affected if the agreement is not signed. When the adjudicator accepted the appointment he entered into a contract with the parties, one of the terms of which was that he and the parties would sign the agreement. The terms therefore apply, although there may be some uncertainty over identity or fees.

6.7.2 The ICE contracts

The ICE contracts and subcontracts incorporate the ICE Adjudication Procedure (1997) which in turn provides that the parties will enter into an appointment on a standard form within seven days of being requested to do so. Once again, this standard agreement contains valuable recitals, and also has a standard form of schedule detailing the adjudicator fee agreement. This is in rather more detail than in the JCT version, and will be considered in more detail in Chapter 8. The terms are again very simple, but differ slightly from the JCT agreement:

(1) The rights and obligations set out in the ICE Adjudication Procedure are incorporated into the agreement
(2) The adjudicator accepts the appointment and agrees to follow the procedure
(3) The parties agree to be jointly and severally responsible for fees and expenses
(4) The parties and the adjudicator agree to maintain the confidentiality of the adjudication
(5) The adjudicator agrees to inform the parties if he intends to destroy documents sent to him in relation to the adjudication and to retain them for a further period if asked to do so.

CHAPTER SEVEN
THE ADJUDICATOR'S DECISION

7.1 The duty to decide

Paragraph 20 of the Scheme sets out the adjudicator's duty to decide the matters in dispute, and also deals with several other issues to be considered in this chapter:

'20. The adjudicator shall decide the matters in dispute. He may take into account any other matters which the parties to the dispute agree should be within the scope of the adjudication or which are matters under the contract which he considers are necessarily connected with the dispute. In particular, he may –
  (a) open up, revise and review any decision taken or any certificate given by any person referred to in the contract unless the contract states the decision or certificate is final and conclusive,
  (b) decide that any of the parties to the dispute is liable to make a payment under the contract (whether in sterling or some other currency) and, subject to section 111(4) of the Act, when that payment is due and the final date for payment,
  (c) having regard to any term of the contract relating to the payment of interest decide the circumstances in which, and the rates at which, and the periods for which simple or compound rates of interest shall be paid.'

Under the Scheme the adjudicator's duty to decide the matters in dispute is unequivocal. His duty is not to use his best endeavours to come to a decision, or to decide the matters in dispute if they are capable of being resolved. He must either decide the matters in dispute or resign under paragraph 9(1). If he resigns because he discovers that there has been a previous reference of the matter to adjudication and a decision has been given, or because he is not competent to decide it because the dispute varies significantly from the dispute referred to him in the referral notice, he is entitled to be paid. He is not entitled to be paid if he resigns for other reasons, including the belief that it is not possible to make a decision.

During the discussions and consultations that led to the production of the Scheme in its current form, there was some talk about what the adjudicator should be permitted to do if he felt that the matters in dispute were not suitable for resolution by the Scheme's adjudication procedures. It was suggested that he should be able to terminate the proceedings or 'decide' that he could not decide. These possibilities did not find their way into the final version of the Scheme, and the adjudicator does not have them at his disposal. We are left with the simple requirement that the adjudicator shall decide the matters in dispute.

There are times when the adjudicator feels that he has an impossible job. This may be because issues arise during the adjudication about which the adjudicator has no experience and with which he believes he is not qualified to deal. Otherwise he may feel that the sheer volume and complexity of material he is being asked to consider, and which he has a duty to consider under paragraph 17, is such that he cannot possibly deal with it within the time available.

When faced with these moments of self-doubt, the adjudicator must remember:

(1) That he is not a judge of the Technology and Construction Court, nor is he an arbitrator. Such tribunals are faced with the responsibility not just of reaching a decision, but of reaching the ultimate decision. Whilst in the post-Woolf era there is a requirement of proportionality in litigation, and the arbitrator has a duty to avoid unnecessary expense, both judges and arbitrators are able to take very much longer in order to explore all aspects of the matter, and hopefully produce the correct decision. It is clearly not possible to deal with as much detail in a 28 day adjudication as can be digested in litigation over two years.

(2) That even a civil court judge or arbitrator is reaching a decision on the balance of probabilities, not on the criminal standard which requires proof beyond reasonable doubt. The adjudicator is not required to go any further. Having read all the documents, asked all the questions that he wishes to ask and is able to ask in the time available, the adjudicator has to decide the result that he thinks is most probably right.

(3) That he has an express power under paragraph 13(f) of the Scheme to appoint experts, assessors or legal advisers, pro-

viding he notifies the parties of his intention to do so. Technical expertise may be very helpful and many adjudicators are appointed because they have specific technical skills, but the real skill of the adjudicator lies in his ability to identify the issues and assess evidence. Just as a judge relies on expert witnesses to explain technical matters to him, the adjudicator can rely on experts appointed directly by him. Similarly if the adjudicator is unclear about the law, he should consult a suitably experienced lawyer. In either case, the decision must remain the adjudicator's.

(4) That he has wide discretion over procedures. If a party's submissions are unnecessarily voluminous, he can direct that they be resubmitted in shorter versions, and he can also direct the use of spreadsheets or other means of data presentation that will assist in identifying issues.

(5) That extensions of time are possible. Whilst the adjudicator cannot give himself an extension of time, the claimant is likely to agree to the initial 14 day extension if the adjudicator is not going to be able properly to consider his case without it. Further extensions, needing the consent of all parties, may be more difficult, but once again the parties may feel that it is not in their interests to insist that the adjudicator produces his decision before he is ready to do so. That said, the adjudicator should not seek an extension of time unless it is really necessary. The immediacy of adjudication is one of its most important attributes.

When these matters are borne in mind, there is really no dispute that an adjudicator cannot decide. He may not be confident that his decision is absolutely right in every detail, but he is not being asked to produce such a decision.

The published rules for adjudication other than the Scheme do not differ materially from the Scheme in this respect.

7.2 *The matters in dispute*

The adjudicator must be entirely clear about what matters are in dispute and therefore what matters he is required to decide. The first place to look to establish this will always be the notice of adjudication, which under paragraph 1(3) of the Scheme should set out 'the nature and a brief description of the dispute' (inter alia). Unfortunately many notices of adjudication are drafted without absolute clarity, and the referral notice that follows a few days later will often refer to other matters as well as or instead of those specified in the notice of adjudication. The respondent will then often wish to include a counterclaim in his response to the referral notice, and hence matters may become confused.

The second sentence of paragraph 20(1) of the Scheme is itself a little confusing. The statement that the adjudicator 'may take into account' matters that are not 'the matters in dispute' suggests that he might be doing something other than deciding those matters. He is certainly under no obligation to consider any matters other than the 'matters in dispute'.

The following is suggested as a guide to what the adjudicator can and cannot do in drawing up a list of matters that he is required to decide:

(1) Matters stated to be in dispute in the notice of adjudication are 'matters in dispute' and the adjudicator is required to decide them.
(2) Matters described in the notice of referral but not identified as matters in dispute in the notice of adjudication may be added to the list of matters to be decided if paragraphs (4) or (5) below apply.
(3) Matters raised as counterclaims, not having been raised as matters in dispute in the notice of adjudication, may be added to the list if paragraphs (4) or (5) below apply.
(4) If the adjudicator on reading the notice of referral or the response realises that matters other than the dispute identified in the notice of adjudication are being raised, he should ask the parties whether they agree that they should be dealt with by him as adjudicator, and a decision given with regard to them. The adjudicator cannot be compelled to deal with such matters.
(5) If the adjudicator forms the view that some other matter under the contract is necessarily connected with the dispute identified in the notice of adjudication, because for example a decision on the matter is a necessary step to arriving at a decision on the matters in dispute, he can take it into account in arriving at his decision.
(6) If the referral notice or response document raises some matter which was not identified in the notice of adjudication, and there is no agreement about whether it should be included and the adjudicator does not form the view that it is necessarily connected with the original dispute, he should not give a decision on it.

A few examples of situations that can arise may be helpful:

(1) The notice of adjudication given by a subcontractor to a main contractor takes the form of a simple letter. In it the subcontractor states that the main contractor has underpaid the subcontractor by £10,000 in the last interim payment. The notice of referral sets out the subcontractor's claimed interim account, with variations valued by the subcontractor, and also a claim for loss and expense arising from the delay and disruption to the subcontract works caused (allegedly) by interference with the subcontract works by the main contractor. The subcontractor says that it is entitled to an extension of time, and that a contra charge of £5000 deducted by the main contractor (after an appropriate notice of intention to withhold) was unjustified. In its response, the main contractor gives an alternative, and rather lower, valuation of variations and states that the subcontractor was late and is not entitled to any extension of time.

The subcontractor's claim set out in its notice of referral is all within the rather brief description of the dispute in the notice of adjudication, with the exception of the claim for an extension of time. The adjudicator should invite the parties to agree that he should deal with that claim. If the main contractor does not agree, the adjudicator may conclude that it is necessary to consider the entitlement to an extension in order to enable him to decide whether the contra charge (which was effectively within the notice of adjudication) is sustainable.

(2) A subcontractor claims for the price of variations. In the notice of adjudication it identifies the variations in dispute and states that it has been underpaid for them by the sum of £25,000. The notice of referral does not introduce any new matter. The main contractor responds by setting out its valuation, but also seeks to counterclaim in respect of allegedly defective work. The defects are in the original work and have nothing to do with the variations.

The adjudicator should consider whether he is prepared to deal with the arguments on defects. He may decide that they will involve very substantial additional time that he will be unable to commit in the 28 day period. If he is prepared to deal with them, he should ask the parties whether they agree to him doing so. The subcontractor may well not agree, but on the other hand he may prefer to have the whole matter resolved by the same adjudicator in order to save unnecessary cost. If he does not agree, the adjudicator should consider whether the defect issue is necessarily connected with the original dispute. The likely answer is that there is no connection, and therefore the adjudicator will not deal with it and will not take it into account.

(3) In a variation of the situation in (2) above, the alleged defects are within the variations that are in dispute. The main contractor says that the value of the variations is therefore reduced. As the original dispute was the value of the variations, this is within that dispute and is a matter that must be decided by the adjudicator. If the notice of adjudication had been even more specific than it was, and had set out the dispute as being one about the principles of how the variations were to be valued, then the answer would have been as in (2).

(4) A main contractor gives notice of adjudication saying that the dispute is about the deduction of liquidated damages by the employer. When it serves its notice of referral it asks not just for the return of the liquidated damages but also for an extension of time, on the basis of late information from the architect. It also raises a claim for loss and expense resulting from the late information.

Once again the adjudicator will first consider whether he is prepared to deal with these additional matters. If he is he will ask the parties whether they agree. The employer will probably refuse. The adjudicator may then decide that the claim for an extension of time is necessarily connected with the dispute about liquidated damages. It is therefore a matter that he will take into account. The claim for loss and expense is also connected, but not 'necessarily connected' and he will therefore not decide that matter.

The JCT series of contracts and subcontracts, the Government Contract Rules for use with GC/Wks/1, and the CEDR Rules are not so obviously helpful to the adjudicator. There is no express provision for the adjudicator to decide that a matter not specified in the notice of adjudication is necessarily connected with the dispute and therefore something that should be taken into account. Nevertheless, if it is something that the adjudicator must consider in order to decide the matter in dispute, it is common sense that he must be able to do so, otherwise he would be unable to come to a decision, or alternatively would be highly likely to come to the wrong decision. As adjudication is a contractual matter it is still open to the parties to agree that additional matters should be within the scope of the adjudication, and of course as the adjudicator's appointment is also contractual, the adjudicator must agree to deal with such additional matters.

The ICE Adjudication Procedure deals formally with the issue, but does no more than set out the position as described above:

'5.2 The Adjudicator shall determine the matters set out in the Notice of Adjudication, together with any other matters which the Parties and the Adjudicator agree should be within the scope of the adjudication.'

The CIC Rules set out a similar provision:

'20. The Adjudicator shall decide the matters set out in the Notice, together with any other matters which the Parties and the Adjudicator agree shall be within the scope of the adjudication.'

The TeCSA Rules give the adjudicator more responsibility for deciding what to decide:

'11. The scope of the Adjudication shall be the matters identified in the notice requiring adjudication, together with
  (i) any further matters which all Parties agree shall be within the scope of the Adjudication, and
  (ii) any further matters which the Adjudicator determines must be included in order that the Adjudication may be effective and/or meaningful'

Under these rules, the adjudicator can determine that in order to make the adjudication 'effective and/or meaningful' an extra matter must be included in the process. By including this extra matter in the scope of the adjudication, the adjudicator is adding it to the list of matters on which he is to give a decision. This is not just taking the extra matter into account when deciding the original dispute. The adjudicator will be giving a decision on this additional matter that will remain binding on the parties until subsequent litigation or arbitration, despite the fact that it was not expressly included in the notice of adjudication

Unlike the Scheme provision, and indeed the other sets of rules, the adjudicator cannot object if the parties decide that an additional matter should be included in the adjudication. By accepting appointment, the adjudicator has agreed to be subject to the TeCSA

Rules, which at paragraph 11(i) give the parties the right to add matters.

On the other hand the adjudicator has been given much greater powers to rule on the scope of the adjudication:

> '12. The Adjudicator may rule upon his own substantive jurisdiction, and as to the scope of the Adjudication.'

It is essential that an adjudicator's decision should be enforceable. This topic is discussed in Chapter 9. As will be seen then, the principle question that has arisen in reported cases regarding enforcement of adjudicators' decisions is that of jurisdiction. This was foreseen by Sir John Dyson in *Macob Civil Engineering Ltd* v. *Morrison Construction Ltd* (February 1999), the first reported case on adjudication. He was not then dealing with a jurisdiction point, but rather with alleged procedural irregularities that he held did not affect the status and enforceability of the adjudicator's decision. 'Different considerations may well apply,' he said, 'if he purports to decide a dispute which was not referred to him.'

7.3 Power to open up certificates etc.

Many standard form construction contracts link entitlement to payment only indirectly to the work carried out, introducing a requirement for certification by an architect, engineer, project manager or similar. When the contractor claims that he is entitled to payment of a particular sum, the simple answer is to ask whether there is a certificate for payment of that sum. If there is no certificate, there is no entitlement.

This situation lay at the heart of the 1984 decision of the Court of Appeal in *Northern Regional Health Authority* v. *Derek Crouch Construction Co*, in which it was held that the court had no powers to open up, revise or review certificates given by the architect in a standard building contract. Such powers were expressly given to arbitrators under the terms of the contract. That decision was heavily criticised but not overturned until the House of Lords gave their decision in *Beaufort Developments (NI) Ltd* v. *Gilbert Ash NI Ltd* (1998). It is now established that the courts have the power to open up architect's certificates, and the logic of the decision suggests that the courts can also deal effectively with similar disputes in engineering contracts.

Whatever the powers of the courts may be, however, it is not at all clear that an adjudicator would have a similar power unless he is expressly given it by the rules of procedure that he is implementing. The Scheme does this through paragraph 20(a).

This power is not completely unfettered. If the contract provides that a decision or certificate is final and conclusive, the adjudicator does not have the power to open it up, revise or review it. This is designed to ensure that provisions such as clause 30.9 of the standard JCT contract (the finality of the final certificate) are not disturbed.

When the Scheme was published there were concerns that this restriction on the adjudicator's ability to open up a certificate might be used by those anxious to restrict the scope of adjudication. A main contractor might provide in its standard conditions of subcontract, for instance, that the decision of its managing surveyor with regard to the value of the subcontract works would be final and conclusive.

In practice, few main contractors or employers appear to have tried to take advantage of this provision to prevent subcontractors and others from exercising their right to go to adjudication. Some commentators who seek to defend the subcontractor from contractual provisions that are perceived to be unfair, argue that a term can be implied that any decision that has contractual effect would be subject to a requirement of reasonableness. The decision in *John Barker Construction Ltd* v. *London Portman Hotel Ltd* (1996) is cited as authority for this. In that case, the contractor challenged the architect's extension of time on the basis that it was not reasonable. The court agreed and it was held that there was an obligation on the architect in a standard JCT contract to act fairly:

> 'I find quite unacceptable the suggestion that the parties can have intended that a decision on a matter of such potential importance should be entrusted to a third person, who was himself an agent of one party, without that person being under any obligation to act fairly. It seems to me to go without saying that the parties must have intended the decision-maker to be under such an obligation, the imposition of which is necessary to give efficacy to the contract.'

It may be appropriate to extend this argument to the situation where an employee of the main contractor apparently has authority to decide the subcontractor's entitlement, so that the decision could be challenged either in litigation or arbitration. It is however difficult to see how the challenge could be brought in adjudication under the Scheme in the face of the express exclusion from the adjudicator's jurisdiction set out in paragraph 20(a).

The JCT system of contracts and subcontracts also provides for the adjudicator to have the ability to open up, review or revise any 'certificate, opinion, decision, requirement or notice issued given or made under the Contract [Sub-Contract] as if no such certificate, opinion, decision, requirement or notice had been issued given or made'. The exceptions for the certificates that are to have final effect are set out in the relevant clauses dealing with the conclusive effect of those certificates. The ICE contracts also contain a similar provision, as do the CIC Rules.

The TeCSA Rules are of course designed to operate with a variety of contract formats. Paragraph 16 of those rules takes an effective shortcut to give the adjudicator appropriate powers under any contract:

> 'The Adjudicator shall have the like power to open up and review any certificates or other things issued or made pursuant to the Contract as would an arbitrator appointed pursuant to the Contract and/or a court.'

CEDR takes a similar approach:

> '7... The Adjudicator shall have the power to review and revise any decision made under the terms of the contract to which the dispute relates except where the contract precludes this.'

7.4 Decision on payments

The majority of disputes referred to adjudication relate to payments. The logical consequence of most adjudicators' decisions therefore is that some money should be paid by one party to another, unless the claim has completely failed. It would be unsatisfactory for the adjudicator to decide, for example, that the proper value of work carried out under the contract is £100,000, being £20,000 more than had already been paid, without going on to say that the balance should be paid. Paragraph 20(b) therefore provides that the adjudicator can decide that a party is liable to make a payment under the contract. He can also go on to say when that payment is due and the final date for payment.

These two points are important for the operation of the provisions of the Act regarding payment. If the adjudicator merely stated that a sum of money was due on the date of the decision, the contractual provisions for payment would then apply. In particular the contract (or the Scheme operating in default) would provide a final date for payment. The paying party might then be able to give a notice of intention to withhold a sum from the money to be paid, possibly requiring another adjudication.

The final date for payment has another significance. If the payment is not made by the final date, the person to whom the money is due can give notice of intention to suspend and seven days later, if not paid, he can suspend work.

It is up to the adjudicator to decide when the money is due and when the final date for payment arises. Depending on the circumstances of the case, it may be appropriate for the adjudicator to state that the money became due on the date when he considers it should have become due under the contract in the normal course. In other words the value of a variation carried out during November became due on the same date as the value of the rest of the work carried out in November. The final date for payment then follows in accordance with the contract, or the Scheme if there is no provision.

This is all subject to section 111(4) of the Act which provides:

'Where an effective notice of intention to withhold payment is given, but on the matter being referred to adjudication it is decided that the whole or part of the amount should be paid, the decision shall be construed as requiring payment not later than –
(a) seven days from the date of the decision, or
(b) the day which apart from the notice would have been the final date for payment, whichever is the later.'

There would seem to be nothing objectionable in the adjudicator deciding that the wrongly deducted money became due on a date earlier than the disputed notice of intention to withhold, and that the final date for payment under the contract was 'the day which apart from the notice would have been the final date for payment', both dates being in the past. Such a decision would therefore be construed as requiring payment seven days from the date of the decision.

None of the other published rules for adjudication deal with the power of the adjudicator to decide that one party is liable to make a payment to another party. It seems that the authors of those rules considered that it was obvious that an adjudicator would include such matters in his decision without being told that it was open to him to do so. The other published rules also fail to mention the need to state when a payment became (or is to become) due, and what is or was the final date for payment. It is however helpful for these details to be stated in all cases when an adjudicator is dealing with payments to be made under the contract.

Some forms of subcontract conditions used by main contractors provide that if the adjudicator should decide that a sum of money is payable to the subcontractor by the main contractor, he should require that sum to be paid into a stakeholder account, or alternatively state that he may require the payment to be made to a stakeholder account at his discretion. This is an echo of the procedure that used to be available in adjudication before the Act in DOM/1 and other subcontracts used in conjunction with the JCT forms of main contract. It is introduced into main contractors' own forms in an attempt to avoid the risk of money being paid to a subcontractor who becomes insolvent before the main contractor has had an opportunity to take the matter to arbitration or litigation and effectively have the decision reversed. In those circumstances the money would be secure and could be repaid to the main contractor.

Such a provision would appear to be contrary to the intention of the Act, in that the subcontractor would not obtain any tangible benefit of the adjudication for potentially a very long time. Nevertheless there is nothing obvious in the Act that seems to prevent such a clause being effective.

The wording of the clause differs from one main contractor's form to another, and the subcontractor's argument against the operation of the clause will of course have to deal with the specific wording. Subcontractors may wish to adopt the following arguments:

(1) To impose a requirement on the adjudicator to order payment to a stakeholder account is an effective restriction on the ability of the adjudicator to make his decision. Because it is a limitation on the adjudication process that is not expressly permitted by the Act, it prevents the adjudication provision of the subcontract from complying with the Act, and therefore renders the adjudication provision non-compliant. The Scheme therefore applies, which gives the adjudicator an unfettered ability to order a payment to be made to a party (under paragraph 20(b)).
(2) Alternatively the adjudicator cannot order the payment to a stakeholder account until he has decided that a sum is due to the subcontractor. Having made that decision, he is *functus officio*, and is no longer able to order anything at all.

(3) The intention of Parliament was clearly to provide for prompt payment of sums found due by an adjudicator. Following the decision of the House of Lords in *Pepper* v. *Hart* (1993) the courts are able to consider Parliamentary materials in establishing the true intent of legislation where the statutory provision is thought by the court to be ambiguous or obscure or where its literal meaning would lead to an absurdity. This purposive approach would be fatal to a stakeholder requirement, but it is by no means certain that the court would consider that the Act is ambiguous or obscure or that to give force to a stakeholder provision would lead to an absurdity.

Some main contractors' legitimate concerns about paying substantial sums to subcontractors pursuant to adjudicators' decisions, with the risk that the subcontractor may be insolvent or at least impecunious and unable to repay when the decision is reversed, may have been reduced by the comments of Sir John Dyson in *Herschel Engineering Ltd* v. *Breen Property Ltd* (April 2000) and the Court of Appeal's decision in *Bouygues UK Ltd* v. *Dahl-Jensen UK Ltd* (July 2000). The comments in *Herschel* suggest that the impecuniosity of the receiving party may influence the court to give a stay of execution of a summary judgment based on an adjudicator's decision, and in *Bouygues* it was said that insolvency set-off rules may apply. These decisions are discussed further in Chapter 9.

If the contract makes no mention of the possibility of an order by the adjudicator that money should be paid into a stakeholder account, it is not open to the adjudicator to make such an order. In *Allied London and Scottish Properties plc* v. *Riverbrae Construction Ltd* (Lord Kingarth, July 1999), an employer resisted enforcement of an adjudicator's decision. It argued that in the light of various claims that it was making against the contractor in other proceedings and had also advanced in the adjudication, the money found due to the contractor should be placed in a secure account. The adjudicator rejected that and said that he had no power to do so. The court agreed:

'Such an order would plainly, in effect, have been to sustain the petitioners' claims to retention which the adjudicator had just rejected. Whatever wide powers may be given to adjudicators to facilitate speedy resolution of the disputes before them, no power is given to make decision contrary to the rights of the parties arising as a matter of law.'

Particular care must be taken by the adjudicator deciding that a payment is to be made by a party to ensure that he deals with VAT. It is a common feature of disputes in the construction industry that the parties express their claims and negotiate settlements in figures net of VAT, and then argue about whether the figures are to be taken as being subject to the addition of VAT or not. Composite claims often include items on which VAT is chargeable, such as the valuation of work, and items on which VAT is not chargeable, such as damages or loss and expense. Unless the adjudicator states that the sum ordered to be paid is to have VAT added to it and paid in addition, the figure set out in the decision will be effectively a gross sum, and no additional VAT will be claimable. The party receiving the payment may still be treated by HM Customs and Excise as having received an element of VAT and will have to account for it.

7.5 Interest

Paragraph 20(c) of the Scheme deals with interest. It states that the adjudicator may:

> 'having regard to any term of the contract relating to the payment of interest decide the circumstances in which, and the rates at which, and the periods for which simple or compound rates of interest shall be paid.'

This paragraph has led to some confusion and debate. Some believe that the adjudicator is empowered by these words to award interest in any adjudication under the Scheme. The paragraph is read as if it said, 'The adjudicator can decide that one party pay interest to the other providing he takes account of any term of the contract relating to interest'. This would suggest that the adjudicator has complete discretion in such matters unless there is a relevant contractual provision.

This is not correct. There is no automatic right to interest on any debt or other entitlement. The courts can award interest only because of statutory authority given to them, and arbitrators take their power to award interest from the Arbitration Act 1996. The Housing Grants, Construction and Regeneration Act does not confer any similar power on adjudicators.

An adjudicator can therefore only decide that interest should be paid if there is a term of the relevant construction contract giving an entitlement to interest. That term might be express, or it might be implied by virtue of the Late Payment of Commercial Debts (Interest) Act 1998. In applying the term of the contract he must of course have proper regard to that term, and if the rate or period or formula for compounding the interest are prescribed by the contract, the adjudicator has no discretion.

The Late Payment of Commercial Debts (Interest) Act 1998 (LPCDA) provides at section 1(1):

> 'It is an implied term in a contract to which this Act applies that any qualifying debt created by the contract carries simple interest subject to and in accordance with this Part [of the Act].'

The LPCDA applies to all construction contracts providing that the employer and contractor (or equivalent) are each acting in the course of a business. 'Business' includes the activities of a government department or local or public authority. The statutory right cannot be excluded unless the contract expressly provides an alternative substantial contractual remedy for late payment of the debt.

Although the LPCDA was passed in 1998 it is not yet fully in force. From 1 November 1998 it applied to contracts for the supply of goods or services (including construction services) if the supplier of the goods or services was a small business, and the purchaser of those services was a large business or a UK public authority. 'Small business' and 'large business' are defined in the Late Payment of Commercial Debts (Interest) Act 1998 (Commencement No. 1) Order 1998. A small business is one that employs 50 or fewer employees, and a large business is one with more than 50 employees. These definitions are subject to complex calculations to take account of the varying status of those who might be considered to be employees, and part-time workers. In any case where the size of business is in doubt, detailed reference must be made to the schedules to the commencement order. By further regulations (the Late Payment of Commercial Debts (Interest) Act 1998 (Transitional Provisions) Regulations 1998, a business is presumed to be large unless the contrary is proved. Interest is payable at the rate of 8% over the official dealing rate of the Bank of England (otherwise known as base rate).

There have been two further commencement orders bringing the LPCDA into force in contracts between small businesses and an extended list of public authorities, but contracts between large businesses, or contracts involving supply from large businesses to small businesses are not yet included.

7.6 Form and content of the decision

7.6.1 Preliminary matters

Unlike the Arbitration Act 1996, which states that an arbitrator's award shall be in writing and specifies other requirements, the Act does not mention any formalities required to be included in the decision of an adjudicator. Whilst we have seen that paragraph 20 of the Scheme gives the adjudicator power to include various matters in his decision, there is nothing in the Scheme that states that the adjudicator shall include anything at all. The JCT series of contracts and subcontracts and the TeCSA Rules require that the decision shall be in writing, but no other requirement is stated in any of the published rules and standard forms. The adjudicator is left to decide for himself how he is to communicate his decision to the parties, and what form the decision should take.

The adjudicator will be trying to achieve a number of objectives in setting out his decision:

(1) He will wish to achieve clarity
(2) In particular, he will wish to make it absolutely clear who is to do what, and by when
(3) He will wish to ensure that his decision is effective and can be enforced if necessary
(4) He will wish to avoid any suspicion that he has not been impartial or has failed to take account of what has been submitted to him by either party
(5) He will wish to avoid giving any party a valid reason for not paying him.

Before starting to write his decision, the adjudicator will wish to revisit the contract to ensure that he is in a position to do all that he is required to do. Whilst standard procedural rules may have been incorporated either expressly, or in the case of the Scheme by default, there may be some specific requirement set out in the contract. If he does not comply with that requirement, it is possible that his decision will not be treated as a decision at all, although this is unlikely in view of the words of Sir John Dyson in *Macob Civil Engineering Ltd* v. *Morrison Construction Ltd* discussed earlier. Equally serious from the adjudicator's point of view is that if he does not deal with the matter as he has impliedly agreed to do, he may not be entitled to his fee.

Having reassured himself as to any specific requirements, the first question, unless he is acting under the JCT system or TeCSA Rules, will be whether the decision should be in writing. This will not normally be difficult to answer. It may be appropriate in some adjudications to deliver a decision orally at a meeting, but even when this has been done it will be prudent to confirm the decision in writing immediately thereafter. Enforcement of an oral decision would require evidence of what the decision was, and if it was not given or confirmed in writing by the adjudicator there is risk of serious argument that would effectively prevent enforcement without a full trial or arbitration.

The adjudicator will almost certainly wish to sign and date his decision, to distinguish the final document from any previous draft, and to avoid any argument about whether the document is his decision or that of someone else.

Still bearing in mind the potential need for the successful party to enforce the decision, the adjudicator may wish to include a recital setting out the details of the contract under which the dispute arose. The court may wish to consider whether the contract was a construction contract at all, whether or not it was in writing and whether it was formed before or after 1 May 1998.

The decision will have no effect unless the adjudicator was properly appointed. He may therefore wish to recite the agreement of the parties to appoint him, or alternatively the process by which he was nominated by an adjudicator nominating body.

Whilst the court is unlikely to decline to enforce a decision because of procedural irregularity, the adjudicator may wish to record the procedures that have been adopted. If he sets out a chronology of the adjudication, stating when meetings were held, who attended and their purpose, and also lists when various documents were received, he can demonstrate that the parties have had their chance to put their arguments. He is also making it clear that all the submissions received have been considered. Particularly important from his own point of view, he can record when the 28 days started to run and what extensions of time if any were agreed, and therefore can demonstrate that he has produced his decision on time.

The importance of identifying the matters in dispute that the adjudicator has to decide has been considered earlier in this chapter. The adjudicator should set out concisely what matters were within the scope of the adjudication, either because they were stated in the notice of adjudication, because the parties have agreed to include them, or because he has exercised a discretion such as that contained in paragraph 20 of the Scheme. If matters have been added he should make it clear when and how agreement was reached or his discretion was exercised. This will assist in avoiding arguments about jurisdiction in any application to enforce the decision.

7.6.2 The decision itself

Having stated clearly what the matters were on which he is to decide, he should proceed to state what his decision is on each matter. Each decision should be concise, and he should avoid including any comment in the nature of reasons for his decision. If reasons are to be given, they will be set out in a separate section of the document, either before or after the decision itself. If the question put to him was the value of a specific item of work, he should state what value he has decided should be attributed to it. If he has been asked to decide on the appropriate extension of time in respect of a particular matter, he should simply state what extension of time, if any, the contractor is entitled to be given. If the consequence of that decision is that a sum of money should be paid by one party to another, that consequence should be expressed in a further paragraph. Absolute clarity is required in giving the decision on each point. If the decision is not absolutely clear there will be doubt as to what is supposed to happen as a result of the adjudication, and there will be an increased chance of an argument that the adjudicator has exceeded his jurisdiction and has decided something that he was not asked to decide.

Having stated his decision on the points in the adjudication, he should state, if appropriate, that one party is to make a payment to another, and if so, how much. This is expressly suggested by paragraph 20(b) of the Scheme and is implicit in the other published rules. Again, absolute clarity is required. As discussed earlier in this chapter, he should deal clearly with VAT and, if appropriate, interest.

The adjudicator should then deal with the time for payment. This involves not just the date the payment became due and the final date for payment, which affect the ability of the paying party to give a notice of intention to withhold and the receiving party's right to suspend work, but also the date by which the adjudicator requires the parties to perform in accordance with the decision. Paragraph 21 of the Scheme deals with the time for performance if the decision is silent:

'21. In the absence of any directions by the adjudicator relating to the time for performance of his decision, the parties shall be required to comply with any decision of the adjudicator immediately on delivery of the decision to the parties in accordance with paragraph 19(3).'

It may seem potentially confusing to have various dates for payments to be due, finally due and made in accordance with the adjudicator's directions, but the position is logical. The adjudicator may well decide on 1 December that a payment became due under the contract on 31 October, the final date for payment of that sum was therefore 17 November, and he directs that it be paid forthwith.

The question of costs, both his own remuneration and the costs of the parties, is considered in Chapter 8. The adjudicator should express his decision on these matters clearly.

Paragraph 23 of the Scheme states that the adjudicator may order any of the parties to comply peremptorily with his decision or any part of it. This provision, and paragraph 24 which applies section 42 of the Arbitration Act 1996, is concerned with enforcement and will be considered in depth in Chapter 9. It should be noted here, though, that if the adjudicator wishes to make such an order, he should do so in the decision document.

7.6.3 Reasons

The adjudicator will then have to decide whether or not to give reasons. There has been considerable debate about whether reasons should be given in adjudication. When statutory adjudication was being discussed, and before it came into operation, many commentators argued strongly that reasons should not be given. Several practising adjudicators who adopted that view at the time now say that they will normally give reasons.

Obviously, the parties will understand the decision better if they are told the adjudicator's reasons. This may help the loser to accept the decision and may discourage him from incurring further costs by pursuing the case to litigation or arbitration. It may make it easier to manage the remainder of the contract, if it is continuing, avoiding other disputes in the future. If enforcement proceedings are necessary, the court or arbitrator dealing with the matter will be able to consider the basis for the decision, and be better able to judge whether it was correct. It is also suggested that if the adjudicator is required to give reasons his whole thought process will be more rigorous and more dependable.

These arguments are not as strong as they may appear. The adjudicator should approach the preparation of his decision in precisely the same way whether or not reasons are required. He should always feel able to justify his decision. There is an argument against reasons that there will be a substantial saving in time and cost if reasons are not requested, but that also has little merit because the adjudicator should put as much time and effort into reaching his decision in any event. The time and therefore cost of setting down his reasons on paper are relatively minor compared to the time and cost of arriving at the decision.

The suggestion that a court may be interested in the reasons behind the decision has been shown to be unfounded. The approach of the courts to enforcement will be considered in detail in Chapter 9, but it is clear from comments such as those of Sir John Dyson in *Bouygues UK Ltd* v. *Dahl-Jensen UK Ltd* (November 1999 and upheld in the Court of Appeal in July 2000) that adjudicators' decisions will be enforceable 'pending final determination of disputes by arbitration, litigation or agreement, whether those decisions are wrong in point of law or fact'. The court will clearly be interested to know that the adjudication was properly founded and that the adjudicator had not decided something that had not been put to him for decision, but if satisfied on those points, the court will only wish to know what the decision is, not the reasons for it.

Whilst the parties may understand the decision better if they are given reasons, discouraging them from going on with further litigation or arbitration, close analysis of the reasons may lead them to restate their case in a way which would have more chance of success before a different tribunal. Further litigation may be encouraged rather than avoided.

If the adjudicator has any choice in the matter, he should consider whether in his judgement the parties' interests are best served by being given reasons. If he concludes that they would be better off without reasons he should also bear in mind that he is being paid by the parties and if they require reasons there is at least an argument that their wishes should be respected.

He may still decide that it would be inappropriate to give reasons. In an adjudication about the value of work to be paid on an interim account under a JCT contract, the contractor served notice of adjudication stating that the matter in dispute was the sum payable on that account. The notice of referral then set out the account in detail in order to justify the total sum being claimed. The employer then responded with its version of the account. The valuation of a substantial number of variations was in issue. The adjudicator had to form a view on each variation in order to arrive at his decision as to the correct value of the interim account. His decision was effectively a single sum. He declined to give reasons because he did not think that the reasons would be helpful to the parties. Whilst his decision was enforceable, the value of each individual item, which comprised the reasons, did not form part of the decision. The employer would have been at liberty to ignore the value of each variation in the subsequent interim valuations. Had the notice of adjudication raised the issue of the valuation of individual variations the decision with regard to those items would have been binding and therefore of use in subsequent accounts. It might then have been appropriate to give reasons.

These considerations assume that the adjudicator has a choice in the matter. He may have no such choice. Paragraph 22 of the Scheme provides:

'If requested by one of the parties to the dispute, the adjudicator shall provide reasons for his decision.'

Clause 59 of GC/Wks/1 includes a similar provision.

Where the contract or rules require the adjudicator to give reasons on request, the question arises as to when the request must be made. It is helpful if the adjudicator knows before he prepares his decision whether or not reasons are required. This should not affect in any way the substance of the decision, but it will affect the time that he spends in preparing it, and thus it will affect the quantification of his fees. He will probably include in his decision a determination of the fees payable, and an apportionment of how the fees and expenses are to be borne by the parties.

Although an adjudicator is not an arbitrator, it is generally accepted that the delivery of a valid decision has the same effect as the delivery of a valid award in arbitration. His job is complete and he has no further authority. He is said to be *functus officio*. He is therefore unable to do anything further unless the rules under which he was appointed expressly or impliedly provide for some further action. He cannot then be compelled to produce reasons, if indeed he could have been compelled by a request made before the delivery of the decision itself.

The doubt about when the request for reasons can be made is resolved in GC/Wks/1 by the second sentence of clause 59(10):

'Such requests may only be made within 14 Days of the decision being notified to the requesting party.'

Clearly under this contract the request may be made after the decision has been delivered, and the adjudicator is obliged to provide reasons if asked within the 14 day period.

If he is acting under the New Engineering and Construction Contract, the parties do not need to ask, as clause 90.10 states that he is to provide reasons to the parties and to the project manager with his decision.

The TeCSA Rules take the opposite position. Paragraph 27 states that 'All decisions shall be in writing, but shall not include any reasons.'

Other published sets of rules are permissive, leaving it up to the adjudicator to decide. The JCT system says that the adjudicator shall not be obliged to give reasons for his decision, but does not say that he shall not give them if he wishes to do so. The ICE Adjudication Procedure states at paragraph 6.1 that the adjudicator shall not be required to give reasons, but there is no prohibition. The CIC takes the same line. CEDR makes the permissive position entirely clear:

'9... The Adjudicator may, but shall not be obliged to, give reasons for the decision...'

If the rules to which the adjudicator is working make the position clear one way or the other, the adjudicator must abide by them. If the parties have agreed, as under the TeCSA Rules, that reasons shall not be given, it would be a breach of the adjudicator's contract to disclose his reasons. One or other of the parties may object strongly to such disclosure, which might have a profound effect on the way in which the contract is run. The published reasons may also have serious consequences for subsequent litigation or arbitration.

The following is suggested as a logical way for an adjudicator to consider whether or not he should give reasons:

(1) Does the contract or the relevant adjudication procedure give the adjudicator any choice?
(2) If so, is he prepared to give reasons?
(3) Do the parties want him to give reasons? If so, is it apparent why they want him to do so?
(4) If the parties do not want reasons, the adjudicator should not give them.

(5) If the parties do want reasons, the adjudicator should give them unless he feels that there is good reason to believe, despite the parties' wishes, that reasons would be unhelpful to the parties.

Having decided that reasons are to be given, the adjudicator must consider how to set them out. Many adjudicators approach the preparation of reasons for adjudication decisions in the same way as they would as arbitrators giving reasons for arbitration awards. In some cases this will be appropriate, but that is not necessarily the case. Reasons are required in arbitration, at least in part, to enable the parties and the court to consider whether there is any basis for an appeal. As there is no right of appeal from an adjudicator's decision, this is not relevant. If the matter is to go on to litigation or arbitration the proceedings will start from scratch, not from the adjudicator's decision. If the adjudicator is to give reasons, those reasons are entirely for the use of the parties in understanding the decision that has been given, and, hopefully, making further procedures unnecessary.

The adjudicator will therefore need to ask what form the reasons should take in order to be the most use for the parties. In appropriate cases he might ask the parties what they would like. If the dispute has been about a final account with several hundred items, the parties may not wish to know how the adjudicator has arrived at the value of each one, but they may wish to see what figure the adjudicator has given to each item. If the adjudicator has been working from a spreadsheet set up in a form somewhat like a traditional Scott Schedule, the parties may be happy with a copy of that spreadsheet with a final column indicating the value. Alternatively they may wish to be given detailed reasons for the adjudicator's conclusions on specific major or representative items, with no more than final figures on the smaller items and items that fall within the same class as the representatives.

If the adjudication has been largely about a legal issue the parties may wish to see a full legal analysis, more akin to counsel's opinion than to an arbitration award. Arbitrators are trained not to provide this in their awards for fear of being appealed, but in the absence of any right of appeal this does not need to worry adjudicators.

If the parties do not give any indication of the extent of the reasons that they expect to see, the adjudicator will have to explain how he has come to his decision. He should not feel that length and complexity are virtues in the production of reasons. The parties simply want to know why he has reached his decision. They will also have in mind that they are paying by the hour, and will not wish to pay for unnecessary hours producing many pages of reasons. Some adjudicators' decisions have run into over 100 pages even in relatively minor disputes, and the expense to the parties has been absurd. The adjudicator will again remind himself that he is dealing with the issues in the notice of adjudication, perhaps extended by agreement or by the exercise of his discretion, and will confine himself to dealing with those points.

If the adjudicator has consulted technical or legal experts he will probably be under no obligation to disclose the advice that he has received. Only the CIC Rules require him to do so. Nevertheless the adjudicator should consider summarising any relevant advice that he has received in his reasons.

It does not matter whether the reasons given by the adjudicator appear in the document before the statement of the decisions themselves, or after the decisions. A distinction has always been drawn in arbitration awards between reasons that form part of the award, and reasons that are given in an annex, stated not to form part of the award. Such distinctions are not relevant to adjudicators' decisions.

7.7 Binding nature of the decision

Section 108 of the Act provides:

'The contract shall provide that the decision of the adjudicator is binding until the dispute is finally determined by legal proceedings, by arbitration (if the contract provides for arbitration or the parties otherwise agree to arbitration) or by agreement.

The parties may agree to accept the decision of the adjudicator as finally determining the dispute.'

Paragraph 23(2) of the Scheme echoes that section, and adds an obligation to comply:

'The decision of the adjudicator shall be binding on the parties, and they shall comply with it until the dispute is finally determined by legal proceedings, by arbitration (if the contract provides for arbitration or the parties otherwise agree to arbitration) or by agreement between the parties.'

Adjudication is a contractual process, and the decision of the adjudicator is effective only because the contract says that it will be treated in this way. The Act does not give the adjudicator's decision its binding nature; that comes from the contract. The part played by the statute is to provide that if the contract does not provide that the decision is binding, the Scheme will apply to the contract. Having had the Scheme applied in that way, the contract now provides that the decision will be binding.

The courts have made it clear that they will support the adjudication process by giving the words of paragraph 23(2) their natural meaning. The reported cases on the binding nature of decisions have resulted from resisted applications to enforce those decisions, and will be considered in depth in Chapter 9, but the following words of Sir John Dyson in *Macob Civil Engineering Ltd* v. *Morrison Construction Ltd* summarise the courts' approach:

'Parliament has not abolished arbitration and litigation of construction disputes. It has merely introduced an intervening provisional stage in the dispute resolution process. Crucially, it has made it clear that decisions of adjudicators are binding and are to be complied with until the dispute is finally resolved.'

If a dispute previously referred to adjudication is subsequently the subject of litigation or arbitration, the new tribunal will not be dealing with an appeal from the adjudicator – a completely new process will be started, in the same way as it would have been prior to the introduction of adjudication. The adjudication may have helped to refine issues, and no doubt will affect the way in which the parties present and argue their respective case, but otherwise it will have no effect on the process.

The Act and the Scheme also refer to the possibility of the dispute being concluded by agreement. This did not need statutory authority. A contractual dispute can of course be settled finally by agreement.

The Act also states that the parties may agree to accept the decision of the adjudicator as finally determining the dispute. Consideration has been given (in Chapter 3) to the question of whether an agreement of this type prior to the adjudication effectively converts the process into arbitration. The Scheme does not deal with such a possibility but that does not mean an agreement in such terms is not permissible.

The JCT series of contracts and subcontracts include a provision that the decision of the adjudicator is binding until the dispute is finally determined, but suggest that if the final determination is to be achieved through agreement, that agreement must be made after the decision of the adjudicator has been given. This would seem to be an attempt to preclude the possibility of an agreement in advance to accept the decision of the adjudicator as being conclusive. It would still be open to the parties to make such an agreement, effectively amending the JCT standard terms.

Clause 59 of GC/Wks/1 goes a little further. In order to avoid any possible misunderstanding, it states at clause 59(7) that 'the parties do not agree to accept the decision of the adjudicator as finally determining the dispute'. Once again, if the parties subsequently decided that they did wish to accept the decision as conclusive, before or after the adjudication, they could do so.

The other published procedures merely restate the agreement that the decision shall be binding, required by the Act, and do not attempt to deal with whether or not the decision shall be agreed as being conclusive.

It should be remembered that the adjudicator's decision is binding only on the parties to the adjudication. Unless third parties are obliged by the terms of their contracts with the adjudicating parties to accept the decision of the adjudicator as binding on them, the adjudicator's decision will not be relevant to them. Circumstances in which the decision might be important to a third party include management contracts or other 'cost plus' contracts where the contractor is entitled to recover from the employer whatever he is obliged to pay to the subcontractor, together with a percentage addition or management charge. In a conventional contract, the fact that the subcontractor has been successful in an adjudication in recovering a sum from the main contractor does not mean that the main contractor will necessarily succeed in recovering a related sum from the employer or from another subcontractor.

7.8 Mistakes

Neither the Act nor the Scheme say anything about what is to happen if the adjudicator makes a mistake in his decision. It is clear from the decision in *Bouygues UK Ltd* v. *Dahl-Jensen UK Ltd* (November 1999 and Court of Appeal July 2000) that a mistake does not invalidate the decision or render it unenforceable.

The Arbitration Act 1996 provides a power to the arbitrator, subject to agreement of the parties to the contrary, to correct an award so as to remove any clerical mistake or error arising from any accidental slip or omission, or clarify or remove an ambiguity in the award (section 57(3)).

The ICE Adjudication Procedure contains a similar provision:

'6.9 The Adjudicator may on his own initiative, or at the request of either Party, correct a decision so as to remove any clerical mistake, error or ambiguity provided that the initiative is taken, or the request is made within 14 days of the notification of the decision to the Parties. The Adjudicator shall make his correction within 7 days of any request by a Party.'

Adjudicators were initially very unsure about their position with regard to the correction of clerical errors and the like. The adjudicator in *Bouygues* had expressly reserved the right to correct any such errors, although when asked to do so he had confirmed the erroneous decision. There was doubt about whether he had any power to reserve such a right in any event.

In practice, adjudicators who had clerical errors brought to their attention would in some cases re-issue their decisions with the corrections clearly marked, but without any confidence as to the fate of the corrections if challenged. There was concern that they were by then *functus officio,* and as there was no express power to correct accidental slips or omissions the correction might be of no legal effect.

This uncertainty was removed by the decision of Judge Toulmin in *Bloor Construction (UK) Ltd* v. *Bowmer & Kirkland (London) Ltd* (April 2000). The adjudicator had issued his decision stating that Bowmer & Kirkland should pay Bloor £122,099. In arriving at that decision he had omitted to give credit for sums paid on account. This was pointed out to him and he immediately corrected it. Nevertheless Bloor sought to enforce the original decision, saying that he had no power to make a correction. The judge found that there was an implied term in the contract between the parties and the adjudicator that he should have power 'to correct an error arising from an accidental error or omission or to clarify or remove any ambiguity in the decision which he has reached, providing this is done within a reasonable time and without prejudicing the other party'.

The decision was approved shortly after it had been given by Mr Justice Dyson in *Edmund Nuttall Ltd* v. *Sevenoaks District Council* (April 2000). The adjudicator failed to give credit in his decision for sums already paid by the employer. The error having been brought to his attention, the adjudicator wrote acknowledging the mistake and stating the correct figure to be paid. He said in his letter that he did not believe that he had jurisdiction to correct the error. When the Council made the payment they deducted both the sum previously paid, the subject of the adjudicator's letter, and also deducted liquidated damages. In the following enforcement action the Council argued that there were implied terms in the contract enabling the adjudicator to correct his decision, and also authorising deduction of liquidated damages. Sir John Dyson accepted that there was an arguable case that there was an implied term concerning correction of errors, as found in *Bloor*, but not regarding the deduction of liquidated damages in the absence of the proper operation of the contract machinery.

The limitations imposed by the judge in *Bloor* should be noted. In particular, the correction must be made within a reasonable time. Under the Arbitration Act the correction can be made within 28 days. This would probably be too long within the context of the adjudication timetables. It may even be arguable that the 14 days allowed by the ICE Procedure is too generous.

It should also be noted that this power is limited to genuine slips. The judge in *Bloor* quoted the words of Sir John Donaldson MR from *R v. Cripps (ex parte Muldoon)* (1984):

'It is the distinction between having second thoughts or intentions and correcting an award to give true effect to first thoughts or intentions which creates the problem.'

7.9 Adjudicator's immunity

Section 108(4) states that it is a requirement of every construction contract that the adjudicator should not be liable for his actions as adjudicator:

'108–(4) The contract shall also provide that the adjudicator is not liable for anything done or omitted in the discharge or purported discharge of his functions as adjudicator unless the act or omission is in bad faith, and that any employee or agent of the adjudicator is similarly protected from liability.'

The adjudicator's immunity does not flow directly from this section. The immunity is a provision of the contract between the parties and the adjudicator, but if the contract does not itself provide such immunity, the adjudication provisions of the Scheme will apply.

Paragraph 26 of the Scheme restates the statutory requirement verbatim:

> 'The adjudicator shall not be liable for anything done or omitted in the discharge or purported discharge of his functions as adjudicator unless the act or omission is in bad faith, and any employee or agent of the adjudicator is similarly protected from liability.'

The other published sets of procedure also include an immunity clause in similar but not identical terms. The TeCSA Rules and CEDR Rules include immunity for themselves as nominating bodies. The CIC Procedure goes a little further than the statutory provision:

> '33. The Adjudicator is not liable for anything done or omitted in the discharge or purported discharge of his functions as Adjudicator (whether in negligence or otherwise) unless the act or omission is in bad faith, and any employee or agent of the Adjudicator is similarly protected from liability.'

The addition of the words 'whether in negligence or otherwise' raises a concern about the efficacy of the provisions contained in the other published rules, and indeed the Scheme itself. It is well established that a contractual exclusion of liability will only be effective in excluding liability for negligence if the intention to do so is made clear:

> 'It is, however, a fundamental consideration in the construction of contracts of this kind that it is inherently improbable that one party to the contract should intend to absolve the other party from the consequences of the latter's own negligence. The intention to do so must therefore be made perfectly clear, for otherwise the court will conclude that the exempted party was only intended to be free from liability in respect of damage occasioned by causes other than negligence for which he is answerable.'
>
> (Buckley LJ in *Gillespie Brothers & Co Ltd* v. *Roy Bowles Transport Ltd* (1972))

Section 29 of the Arbitration Act 1996 contains an immunity for arbitrators expressed in similar terms to that required by the Act for the benefit of adjudicators, but arbitrators are able to rely on the Arbitration Act to give them statutory immunity. Adjudicators do not have statutory immunity but merely the benefit of a contractual exclusion clause, which will be construed on the same basis as any other contractual exclusion clause. Unless acting under the CIC Rules or on terms and conditions drafted with this particular concern in mind, adjudicators will have to hope that if a claim is made against them in negligence, a court will conclude that it was the intention of the parties (and not just Parliament) that liability in negligence should be excluded.

There is another potential liability for adjudicators that is not covered by the immunity clause required by the Act. This is liability to third parties arising out of the decision as of course the immunity is a creature of the contract between the disputing parties and the adjudicator which does not involve such third parties. It is difficult to see how a decision that one party owes a sum of money to another party might cause damage to a third party that would be recoverable in an action in negligence by that third party against the adjudicator, but the law of negligence is always developing and it is not possible to say that such a situation would never give rise to potential action. It is more likely that the adjudicator might expose himself to potential action in a dispute regarding quality of work. If the adjudicator is asked to decide whether a structure complies with a contractual specification and he negligently decides that it does, it is not beyond the bounds of possibility that a third party injured in the subsequent collapse of the structure would have a cause of action against the adjudicator.

Once again the CIC Procedure is alone in recognising the potential for such an action. It includes the following:

> '34. The Adjudicator is appointed to determine the dispute or difference between the Parties and his decision may not be relied upon by third parties, to whom he shall owe no duty of care.'

Unfortunately this may not be enough to prevent a third party from relying on the adjudicator's decision, or from arguing in litigation that he had relied on it. It is possible that when faced with such a claim for the first time, the courts might decide that in the circumstances of adjudication the adjudicator does not owe a duty of care to third parties.

It is theoretically possible for the adjudicator to obtain contractual indemnities from the parties in respect of potential liabilities to third parties, but given the speed of the appointment process in most cases, it is unlikely that such a provision could be successfully negotiated in practice. Even if such an indemnity is obtained, there is always a risk that the indemnifying party or parties will be unable to provide an effective indemnity when the time comes to call for it.

Given the uncertainties surrounding the potential liability of adjudicators it is important that adjudicators should maintain suitable professional indemnity insurance cover. Indeed it is a requirement of some adjudicator nominating bodies that any adjudicator appointed by them have appropriate cover. Before appointing an adjudicator to any dispute the RICS requests specific confirmation that such cover is in place.

There is no doubt that the adjudicator can be liable to a party or to all parties if he acts in bad faith, although quite what bad faith really is may be unclear. It has been described as 'malice or knowledge of absence of power to make the decision in question' (*Melton Medes Ltd & Another* v. *Securities and Investment Board* 1995).

CHAPTER EIGHT
COSTS

The Act is silent about the adjudicator's fees and how costs should be dealt with between the parties and by the adjudicator. The parties are therefore free to agree anything they wish about these aspects.

8.1 Adjudicator's right to fees and the power to apportion

The parties may have negotiated or agreed particular terms with regard to the adjudicator's fees when they appointed him. If they have agreed to appoint him on the basis of one of the standard forms of appointment, the terms of that appointment will of course apply, and if no specific terms are expressly set out either in the particular appointment or the standard form applicable, terms may be implied by the application of the rules of the adjudication system applicable to the dispute.

The Scheme provides that:

> '25. The adjudicator shall be entitled to the payment of such reasonable amount as he may determine by way of fees and expenses reasonably incurred by him. The parties shall be jointly and severally liable for any sum which remains outstanding following the making of any determination on how payment shall be apportioned.'

There are two tests of reasonableness to be considered. The fees and expenses of the adjudicator are only recoverable in so far as they are reasonably incurred. The adjudicator has a very wide discretion in establishing how the adjudication will be conducted and his directions are not subject to any need to be reasonable. If however he decides to spend long hours investigating irrelevant points or disputed items of minimal value the parties may object to paying him for his efforts. Similarly if he decides to travel long distances to interview someone who has little or nothing to say on the matters in dispute it may be said that the cost of the travel is an expense that was not reasonably incurred.

Having satisfied the requirement that the fees and expenses are reasonably incurred, they must still be a reasonable amount. No further guidance is given as to how the adjudicator should calculate his fees or expenses in order to satisfy that criterion. There is a wide variation in charging between adjudicators of different professions and seniority. Most adjudicators charge on the basis of an hourly rate, but some will apply a fixed fee for disputes, particularly those involving relatively small sums, and some will add a surcharge over their normal hourly rate to take account of the urgency and substance of the matter.

Comparisons have been drawn with the entitlements of arbitrators, and the accepted principle that arbitrators should be entitled to charge on the basis briefly described above may well be relevant to the position of adjudicators, but it must be remembered that adjudicators are not arbitrators. Their responsibilities are quite different, with adjudicators working to a much more demanding timetable at very short notice but not being required to produce a decision that is final. Furthermore, as discussed below, adjudicators may not be able to secure payment of their fees in the same way as arbitrators.

With that note of caution in most cases it will be appropriate for the adjudicator to charge on a similar basis to that which he would normally adopt in his normal professional work. If the parties decide, either between themselves or through the agency of an adjudicator nominating body, to appoint an eminent leading counsel specialising in international construction disputes, they must expect to pay fees at the level normally charged by him.

The adjudicator's right to fees arises on the making of his decision. He is not entitled under the Scheme to interim payments. This is not stated expressly, but the Scheme includes paragraph 25 within the section entitled 'Effects of the decision', and in any event the appointment of the adjudicator without an express entitlement to interim payments is in the nature of an 'entire contract' where no obligation to pay arises until the completion of the work (as in the well-known seaman's wages case from 1795, *Cutter* v. *Powell*).

The adjudicator can however become entitled to payment of fees and expenses without producing a decision under the express provisions of paragraphs 9 and 11 of the Scheme. If the adjudicator resigns because he finds that the dispute is the same or substantially the same as one which has previously been referred to adjudication and a decision has been taken in that adjudication, he is obliged to resign by paragraph 9(1). In those circumstances paragraph 9(4) gives him an entitlement to his fees and expenses in the same terms as paragraph 25. He is still obliged to satisfy the two tests of reasonableness discussed above.

Paragraph 9(4) gives the adjudicator a similar right when he resigns in circumstances where 'a dispute varies significantly from the dispute referred to him in the referral notice and for that reason he is not competent to decide it'. It might be suggested that if the adjudicator unreasonably delays taking this decision, the time spent by him on the matter in the meantime will not give rise to 'fees reasonably incurred' and will therefore be irrecoverable.

Paragraph 11(1) of the Scheme gives the parties the power to revoke the appointment of the adjudicator at any time. They may, for example, settle the dispute and not require the adjudicator's services any further. Again, in those circumstances, the adjudicator has a right to be paid, again subject to the two reasonableness tests.

If the adjudicator resigns for any reason other than those covered by paragraph 9(4), or the appointment is revoked due to the default or misconduct of the adjudicator, the adjudicator is not entitled to be paid (paragraph 11(2)).

Paragraph 25 of the Scheme, and also those parts of paragraph 9 that deal with the adjudicator's entitlement to recover fees and expenses, suggest that the parties should be jointly and severally liable to the adjudicator, but the wording is not entirely clear. The joint and several liability appears only to relate to that part of the fees and expenses that 'remains outstanding following the making of any determination on how the payment shall be apportioned'. The adjudicator will normally order apportionment of all his fees and expenses, or alternatively will fail to deal with how they are to be apportioned. If he orders an apportionment, the words of the Scheme might be taken to mean that there is nothing left 'outstanding'. It is assumed by adjudicators that these words mean that if one party fails to pay its share of the fees and expenses (which might of course be 100%), the part not paid is a joint and several liability of both parties. The adjudicator can therefore invoice the parties in accordance with the apportionment and still pursue either party for payment. The better practice when operating under the Scheme would appear to be to invoice the parties in joint names and to give a receipt as to part when a payment is made.

When operating under a current JCT contract or one of the associated subcontracts, the appointment will be subject to the JCT Adjudication Agreement. That agreement deals with the adjudicator's entitlement to fees and expenses at clause 3:

'The Contracting Parties shall be jointly and severally liable to the Adjudicator for his fee as stated in the Schedule hereto for conducting the adjudication and for all expenses reasonably incurred by the Adjudicator as referred to in the Adjudication Provisions.'

The Schedule merely states:

'The lump sum fee is £_____
or
The hourly rate is £_____'

'The Adjudication Provisions' are those set out in the form of contract itself.

Once the Adjudication Agreement has been signed, there is no doubt about how the adjudicator's fee should be calculated, because it will be set out in the Schedule. There may however be some difficulty if the adjudicator is appointed before the Adjudication Agreement is signed. In that case the fee may not have been specified, and the adjudicator will rely on an implied entitlement to be paid a reasonable fee.

The agreement does not say that the fee is limited to that which is reasonably incurred. Arguably therefore the adjudicator is entitled to charge even for wholly unnecessary work. In response, an aggrieved party may be able to argue that a term should be implied into the agreement limiting such an entitlement.

The JCT provisions with regard to apportionment of and liability for the adjudicator's fees and expenses are rather more clear than the position under the Scheme. Clause 3 of the Adjudicator's Agreement states unequivocally that the parties are to be jointly and severally liable. This is repeated in the contracts themselves. The contracts also provide that the adjudicator should state how payment is to be apportioned, and in default it is stated that the fees and expenses should be borne by the parties in equal proportions. As under the Scheme, there is no entitlement to fees or expenses if the adjudicator does not produce a decision, unless the parties have terminated the appointment other than because of a failure to produce a decision within the time-scale or at all.

In completing the Adjudication Agreement, the adjudicator should make it clear that the lump sum or hourly rate in the Schedule is inclusive or exclusive of VAT. If the adjudicator is registered for VAT and inserts a figure without specifying, it will be deemed to be inclusive of VAT.

The CIC Model Adjudication Procedure and its agreement

between the parties and the adjudicator is perhaps more comprehensive and straightforward than most of the varieties available. The procedure states:

'29. The Parties shall be jointly and severally liable for the Adjudicator's fees and expenses, including those of any legal or technical adviser appointed under paragraph 19, but the Adjudicator may direct a Party to pay all or part of the fees and expenses. If he makes no such direction, the Parties shall pay them in equal shares. The party requesting the adjudication shall be liable for the Adjudicator's fees and expenses if the adjudication does not proceed.'

The agreement with the adjudicator then contains an agreement by the parties that they will pay the adjudicator's fees and expenses jointly and severally in accordance with the procedure and with the schedule to the agreement, which provides an hourly rate for the adjudicator, to be fixed by agreement, but with a stated daily maximum. It states whether the fees include or exclude VAT. Once this has been completed the position is clear, but again the difficulty arises that the adjudicator may be appointed before the agreement is drawn up and executed.

The CIC Procedure does not however deal thoroughly with fees if a decision is not produced. It provides that the party requesting the adjudication pays if the adjudication does not proceed, but it also provides, at paragraph 23, that the adjudicator may resign at any time on giving notice in writing. Does this mean that if he does so, the party requesting adjudication has to pay all his fees? The obvious answer is that in those circumstances the adjudicator is not entitled to any fee, and therefore the liability placed on the referring party is in a nil amount. But does that mean that if the adjudicator resigns for a 'valid' reason, a situation dealt with by the Scheme, he is not entitled to be paid? There may need to be heavy reliance on implied terms in any matter which does not run the full course.

The Schedule to the ICE Adjudicator's Agreement is also more specific than the JCT equivalent. In it, the adjudicator specifies the hourly rate that he is to be paid and provides expressly for payment of all disbursements properly made. A list of potential disbursements is set out, although the adjudicator's entitlement is not restricted to that list. It provides for the payment of an appointment fee, discussed in the next section, and it clarifies the position regarding VAT on fees and disbursements. Finally it states that the fees will be due seven days after delivery of invoice (except in the case of the appointment fee), and provides for interest at 5% above base on overdue accounts.

The potential difficulty that the adjudicator may be appointed before the fee is specified in the formal agreement is dealt with by the ICE Procedure at paragraph 3.4, which provides for the adjudicator to be paid a reasonable fee. If therefore the adjudicator seeks to include an hourly rate that is excessive, the parties can insist that it be reduced to a reasonable figure.

The ICE Procedure (at paragraph 6.5) also makes it quite clear that liability for the fees and expenses is joint and several. It also empowers the adjudicator to direct that one party should pay all or a part of those fees and expenses, and states that if there is no such direction they should be paid in equal shares. The procedure does not however provide for payment of fees if the adjudication is halted part-way through.

Under the Government contract GC/Wks/1 the adjudicator is appointed generally, and not necessarily with specific reference to one dispute. His fees and expenses are detailed in the schedule to the model form of adjudicator's appointment. There is no provision that limits the fees or expenses to a reasonable figure, and so if the adjudicator is appointed before the appointment is drawn up and signed the parties will have to rely on an implied term that the fees will be reasonable.

Clause 59 of the contract itself provides that:

'...The adjudicator's decision shall state how the cost of the adjudicator's fee or salary (including overheads) shall be apportioned between the parties...'

This is the only indication that the adjudicator may be remunerated by a salary rather than by an hourly rate or a lump sum. This is a curious provision, and the appointment itself makes no similar suggestion. There is no default statement that if the adjudicator has not apportioned the fees between the parties, liability will be joint and several.

The TeCSA Rules deal with the adjudicator's fees at paragraphs 23–25:

'23. If a party shall request Adjudication, and it is subsequently established that he is not entitled to do so that Party shall be solely responsible for the Adjudicator's fees and expenses.

24. Save as aforesaid, the Parties shall be jointly responsible for the Adjudicator's fees and expenses including those of any spe-

cialist consultant appointed under 19(viii). In his decision, the Adjudicator shall have the discretion to make directions with regard to those fees and expenses. If no such directions are made, the Parties shall bear such fees and expenses in equal shares, and if any Party has paid more than such equal share, that Party shall be entitled to contribution from other Parties accordingly.

25. The adjudicator's fees shall not exceed the rate of £1000 per day or part day, plus expenses and VAT.'

These rules are the only ones to deal specifically with the question of responsibility for fees when the adjudicator has been appointed without jurisdiction, although, as we have seen, the CIC Procedure provides for the situation where the adjudication does not proceed. In a 'no jurisdiction' case, it would be natural for the adjudicator to direct that the party who attempted to appoint him when he had no right to do so would have to pay the fee. The difficulty would then arise in the absence of paragraph 23 or its equivalent that the decision would not be that of an adjudicator, as he cannot have been appointed, and there would be no obvious entitlement to be paid at all.

Adjudicators faced with this difficulty under other sets of rules and procedures, including the Scheme, would have to invoke a quasi-contractual remedy under common law, claiming a quantum meruit. Under TeCSA Rules the adjudicator's position is more certain.

The TeCSA Rules do not however assist in establishing how the fee is to be calculated, save to put a maximum on the fee rate of £1000 per day or part of a day, plus VAT. This figure is substantially less than would normally be charged by some members of the TeCSA panel of adjudicators if required to devote a full day to the adjudication, and often an adjudicator will be obliged to spend much more than normal office hours on the matter over several days. It does not mean that the hourly rate will be a maximum of £1000 divided by a particular number of hours deemed to be representative of one day. The adjudicator's fee will probably be calculated at the adjudicator's normal charging rate for his professional work, subject to the maximum for any one calendar day. It is also likely that the entitlement is impliedly subject to both the tests of reasonableness found in the Scheme and discussed above.

As usual, the adjudicator can decide and direct by whom the fees and expenses are to be paid. If he does not do so they are borne in equal shares, with provision for recovery of a party's excess payment from the other(s). Overall liability is joint between the parties, not joint and several.

The CEDR Rules contain complicated provisions regarding fees. Each party is responsible for paying a fee per day of the adjudication, which refers to the days to be spent in hearings. These are calculated on a sliding scale according to the amount in dispute and can total a substantial figure. In addition the parties each pay a small fee for preparation. These fees are agreed with CEDR when the procedure is initiated. The adjudicator does however have discretion to apportion the liability for the fees between the parties when giving his decision.

Finally, the position should be considered where there is no express agreement as to fees in the appointment of the adjudicator, and no mention either in the procedural rules set out or incorporated in the contract. For example, a set of procedures may have been provided in the construction contract which comply with the requirements of section 108 of the Act, thereby keeping out the Scheme, but which say nothing at all about the adjudicator's fees.

The appointment of the adjudicator, either direct between the parties and the adjudicator or through the agency of the adjudicator nominating body, is a contract to carry out work. If a fee is not stated, the law will imply a term that the adjudicator is entitled to be paid a reasonable fee. On the basis of early arbitration cases (such as *Crampton and Holt* v. *Ridley & Co* (1887) and *Brown* v. *Llandovery Terra Cotta etc. Co Ltd* (1909)), it is suggested that liability for payment of that fee will be joint and probably several.

Whilst the law will imply a term that the adjudicator is entitled to be paid, it cannot be said with any certainty that the law would imply a term that the adjudicator has power to apportion liability for fees between the parties if the contract is silent on the point. The arguments raised in the case relating to the power of an adjudicator to award costs generally are relevant (see below), and unless the parties expressly or impliedly give the adjudicator power, perhaps by both asking in their submissions for such an order to be made, it is likely that the adjudicator has no such power.

8.2 Right to require security for his fees

Arbitrators often require the parties to arbitrations to deposit sums on account of fees. These sums are paid into a secure account and the arbitrator will draw on them to pay fees invoiced as the reference proceeds. A substantial sum will often be required before the final hearing, so that the arbitrator can be confident that his charges will be met.

Arbitrators are also able to secure payment of their fees by exercise of a lien over the award. If no or insufficient funds have been paid in advance, the arbitrator will prepare his award and then notify the parties that it is ready for collection on payment of his fees and expenses. The right to do this is now enshrined in section 56 of the Arbitration Act 1996.

Adjudicators sometimes believe that they are able to secure their fees in similar ways. Some adjudicators may be, but only if the right to do so has been given under their contract of appointment. There is no equivalent in adjudication of section 56 of the Arbitration Act and if an adjudicator wished to establish a general right to a lien, not dependent on an express contractual right, he would have to develop an argument based on the common law predecessors to the statutory lien of arbitrators such as *Re Cooms and Freshfield and Fernley* (1850) and *Roberts v. Eberhardt* (1857).

The contractual right to require payment in advance, or as a condition of delivery of the decision, will either arise from the specific terms of appointment, or from the rules of the relevant adjudication procedure. If the parties have agreed to appoint a particular person as adjudicator and have invited him to accept appointment, he may well have standard terms of appointment that will include provisions of this sort.

If however the adjudicator has been appointed after an application to an adjudicator nominating body, the position can be quite different. Typically, the institution concerned will contact the potential adjudicator by telephone or fax and ask whether he is able to accept appointment. The person approached will say that he is, and the institution will nominate. The contract of appointment is made at that time, as explained in Chapter 4. After that point it is too late for the adjudicator to introduce additional terms of appointment, unless the parties are happy to agree them.

The other possible way for the adjudicator to establish a right to payment in advance is through the procedural rules. Most however do not make any such provision. The Scheme makes no mention of payments in advance; the possibility of the exercise of a lien is ruled out by paragraph 19(3), which states:

> 'As soon as possible after he has reached a decision, the adjudicator shall deliver a copy of that decision to each of the parties to the contract.'

He therefore cannot say to the parties, 'I have now reached my decision, but I am not going to give it to you until you pay my fees.'

The JCT series of contracts and subcontracts have a similar provision. Clause 41A.5.1 of JCT 98 provides that the adjudicator is to send his decision to the parties 'forthwith'. GC/Wks/1 also requires the adjudicator to notify his decision to the parties (and to the project manager etc.) within the standard adjudication time limits.

The TeCSA Rules state that the adjudicator may not require any advance payment for security of his fees. This would seem to rule out a requirement that fees be paid in advance of or as a condition for the delivery of the adjudicator's decision.

The New Engineering and Construction Contract requires the adjudicator to reach his decision within the normal time limits, but says nothing about delivering the decision to the parties. There is therefore no express or implied statement that the adjudicator is not able to withhold his decision until he is paid, but there is equally nothing to support the suggestion that he is entitled to do so.

Other standard published procedures are more favourable to the adjudicator. The ICE Procedure states:

> 'At any time until 7 days before the Adjudicator is due to reach his decision, he may give notice to the Parties that he will deliver it only on full payment of his fees and expenses. Any Party may then pay these costs in order to obtain the decision and recover the other Party's share of the costs in accordance with paragraph 6.5 as a debt due.'

It is not clear from this provision whether the adjudicator is expected to state what his fees are seven days before he is due to reach his decision. Such a requirement is not practicable, as of course he will be spending an uncertain amount of time over the following seven days. There is also a conflict between this clause, apparently entitling him to set up a lien, and clause 6.1 that requires the adjudicator to notify the parties of his decision within the standard time limits. That paragraph says nothing about his entitlement to hold on to his decision until he has been paid.

An adjudicator under the ICE Procedure is also entitled to be paid an appointment fee specified by him that is to be paid by the parties in equal amounts within 14 days of appointment. It is deducted from the final fee and is therefore effectively a payment on account. Any surplus over and above the final fee is refunded.

If the adjudicator is fortunate enough to be working under CEDR Rules or the CIC Model Procedure he will be able to require payment before he releases his decision. The CEDR Rules, paragraph 5, state that:

> 'The Adjudicator shall be entitled to withhold the issue of the decision until payment [of the adjudicator's fees and expenses] has been made in full.'

On the other hand he is unlikely to have to do so because the scale fees and a payment on account of the time spent in preparation will have been paid to CEDR at least two weeks before a hearing takes place.

Paragraph 24 of the CIC procedure provides:

> 'The Adjudicator may withhold delivery of his decision until his fees and expenses have been paid.'

Whilst it should not be difficult to establish whether or not the adjudicator has the right to require advance payment of fees, either as security at the start of the adjudication, or as a condition of release of the decision, there remains a rather more difficult question as to what the parties should do if the adjudicator says that he requires advance payment of either variety when he is not entitled to it. If both parties agree that they will not co-operate there is not too great a problem. The adjudicator can be told that advance payment will not be made, and if he is not prepared to proceed on that basis a replacement adjudicator will be appointed. If the adjudicator was appointed as a result of an institutional nomination the institution can be asked to nominate an adjudicator who is prepared to behave in accordance with the relevant contractual rules.

It is unlikely to be as simple as that, however, because the parties may well not be able to agree anything. If the objection to the adjudicator's proposal is unilateral there will be a serious concern that neither party wishes to be the one that upsets the adjudicator. The claimant may also wish to make rapid progress and will not wish to spend several days arguing about this point.

Practical experience suggests two ways in which this dilemma is resolved. If the point is made, politely but firmly, many adjudicators acknowledge that there is no entitlement to require payment in advance and waive the requirement without argument. Many parties decide though that they do not wish to make the point, relying on their ability to recover any fees that have been paid in advance from the other party if successful, and accepting that if they are not to be successful in the adjudication they are likely to have to pay the adjudicator's fees anyway.

If the adjudicator does have power to apportion, having been given that power by the contract or the relevant procedural rules, the question arises as to how he should apportion his fees and expenses. None of the published procedures gives any assistance with this. Where express authority is given it is stated in wide general terms, and the adjudicator appears to have complete discretion. The approach of the courts to the enforcement of adjudicators' decisions, right or wrong (see for example *Bouygues UK Ltd v. Dahl-Jensen UK Ltd*) suggests that they will be unlikely to interfere with any decision of an adjudicator on this point. The adjudicator must therefore do what he thinks is appropriate in all the circumstances.

This will not be difficult if one party has been totally successful or has won every significant point. The other party will normally be ordered to pay the adjudicator's fee. It is not unusual though for the decision to be split, with one party winning on some aspects and losing on others. If the adjudicator concludes that both parties have been responsible for the dispute arising it may be appropriate to split responsibility equally. The adjudicator should not be afraid to become more sophisticated than this in some circumstances. For example, a party may have won a substantial sum on a simple legal issue that required little argument and effectively took no time at all in a meeting with the adjudicator, whereas it lost on several issues that required substantial evidence and took two days of hearing for the adjudicator to deal with. In those circumstances it may be appropriate to require the overall winning party to pay the majority of the fees or the expense of an assessor brought in to assist the adjudicator. The adjudicator should however avoid trying to split the responsibility in too precise a manner.

8.3 Power to order payment of costs

The Scheme does include any provision about the power of the adjudicator to order that one party be responsible for reimbursing part or all of the other party's costs. It is not unusual, however, for the claimant to include in its referral notice a claim for costs either in a specific sum or with a request that the adjudicator assess them.

A decision of Judge Marshall Evans QC, *John Cothliff Ltd v. Allen Build (North West) Ltd* (July 1999) suggested that it was open to the adjudicator to do this. The contract had not contained an adjudication clause, and therefore the Scheme applied. The claimant had asked the adjudicator to determine the costs. The adjudicator had included the following in his decision:

> 'In their claim the claimants request me to determine the payment of costs of and in the adjudication. Under the Housing Grants, Construction and Regeneration Act 1996 I have the power to do this. Whereas in arbitration it is normal for costs to follow the event, in adjudication under the Scheme I may make my decision based on the behaviour of the parties in attempting to resolve their differences.'

He then went on to state that the respondent should pay 70% of the claimant's costs, and postponed assessment of them. The respondent refused to pay the costs and the claimant made an application to the court for summary judgment.

The claimant's argument was based on the decision in *Macob Civil Engineering Ltd* v. *Morrison Construction Ltd* (February 1999) in which Sir John Dyson had held that an adjudicator's decision should be enforced despite disputes about procedural irregularity. In any event, said the claimant, the adjudicator had power to award costs under the Scheme, and therefore there had been no procedural irregularity. The respondent said that the matter was more than procedural irregularity – the adjudicator had no right to deal with costs at all.

The judge decided that the adjudicator did have the power to deal with costs, in particular where an application for costs had been made during the course of the adjudication, and representation had been allowed on the point by lawyers on one side and by claims consultants ('leaders in that specialised field of extracting money from contractors up the line, or it may be denying it to contractors down the line'). The power was analogous to the other powers given to the adjudicator to make directions for the conduct of the adjudication. The fact that the construction contract was a substantial matter and not just 'putting in a window in place of one that was rotten' was also significant. The power to deal with costs was to be implied as a necessary term of the contract to give business efficacy.

This decision was greeted with some surprise by many commentators. Understandably, however, it was seized upon by claimants who now felt able to claim costs in the same way as they would be claimed in arbitration or litigation.

The question of costs was raised again in *Northern Developments*

(Cumbria) Ltd v. *J. & J. Nichol* (Judge Bowsher QC, January 2000). A steel frame subcontractor claimed money due for its works under a DOM/2 subcontract, and the main contractor sought to set off claims for defective work and delays. The relevant edition of DOM/2 did not contain the amendments to include adjudication provisions, and therefore the Scheme applied. Both the referral notice and the main contractor's response asked the adjudicator to make an order for costs against the other party.

The subcontractor was successful in the adjudication, and the adjudicator made an order for costs in the subcontractor's favour. The main contractor declined to pay and started an action seeking declarations that the adjudication decision was null and void and ought not to be enforced. The subcontractor applied for summary judgment.

Having found in favour of the subcontractor on the main issues, Judge Bowsher dealt with costs and in doing so reviewed the decision of Judge Evans in *John Cothliff*. He disagreed with Judge Evans in describing the power to award costs as being analogous to the powers given by the Scheme to give directions for the management of the case, and held that there was no implied statutory power granted to the adjudicator to award costs.

Nevertheless, the parties were able to enlarge the power given to the adjudicator by the Scheme if they wished to do so. In the *Northern Developments* case the question was whether the parties had agreed that the adjudicator should have power to deal with costs. Judge Bowsher said:

'I think that there was such an agreement. One party was represented by experienced solicitors: the other party was represented by experienced claims consultants. Both asked in writing for their costs. Neither submitted to the Adjudicator that he had no jurisdiction to award costs. It would have been open to either party to say to the Adjudicator, I have only asked for costs in case you decide that you have jurisdiction to award them but I submit that you have no jurisdiction to make such an award.

In general, an Adjudicator has no jurisdiction to decide that one party's costs of the adjudication be paid by the other party, but in the circumstances of this case, I find that he was granted such jurisdiction by implied agreement of the parties.'

It is therefore possible to give the adjudicator acting under the Scheme power to deal with costs simply by including claims for costs in the submissions made by the parties. Presumably a claim for costs by one side only would not be sufficient, but it may be possible to find a basis for agreement elsewhere, such as in the oral submissions made to the adjudicator. If there is no material from which the adjudicator can conclude that there has been agreement, the 'general' position described by Judge Bowsher will apply, and there will be no power.

The adjudication procedures incorporated in the JCT series of contracts and subcontracts do not make any mention of a power for the adjudicator to award costs, and the position would therefore seem to be the same as described above as applying to adjudications under the Scheme. The position under the CEDR Adjudication Procedure is the same.

The ICE Adjudication Procedure states unequivocally that the parties should bear their own costs and expenses incurred in the adjudication. The parties could still agree to vary that term and give the adjudicator power to award costs, but the agreement would have to be clearly expressed. The same is true under the TeCSA Rules, which state at clause 21(v) that the adjudicator may not require any party to pay or make a contribution to the legal costs of another party. The CIC Procedure also states that the parties shall bear their own costs and expenses incurred in the adjudication (paragraph 28).

GC/Wks/1 takes a different line. Clause 59(5) provides:

'The adjudicator's decision shall state ... whether one party is to bear the whole or part of the reasonable legal and other costs and expenses of the other, relating to the adjudication.'

There is a problem of timing for an adjudicator who decides that he has power to deal with costs. He will not be in a position to deal with costs until he has decided the principal points in issue in the adjudication. If he is to order that one party is to pay another party's costs, he must do so at the same time as he delivers his decision, and he will need to quantify the costs at the same time. Once his decision has been delivered he is *functus officio* and he cannot do anything further. In practice, he will have to decide what order he is going to make and ask for submissions with regard to the sums involved before he delivers his decision. Alternatively he will have to obtain the express agreement of the parties to extend his authority to enable him to assess costs after the decision has been delivered.

If the adjudicator decides that he does have authority to deal with costs, he has no guidance as to how he should proceed. He will probably decide to award costs on the same principles as an arbitrator, so that broadly speaking the successful party will be awarded his costs unless he has done something that has caused unnecessary expense or in some other way has been unreasonable. He may be persuaded to consider a *Calderbank* type of offer, in which a respondent made a proposal to settle on grounds that are found to be reasonable.

If the adjudicator goes on to deal with the quantum of costs he will again be acting without any formal guidelines. Once again he may approach the exercise in the same way as an arbitrator. He will not be obliged to call for a detailed account as if the costs were being formally assessed (or 'taxed') by a court, and he will probably be able to form a view without difficulty as to the reasonableness of the costs claimed in respect of the relatively short process of adjudication. As with all matters in dispute in the adjudication, the adjudicator should act fairly and allow the paying party a reasonable opportunity to make representations.

CHAPTER NINE
ENFORCEMENT

9.1 The Act and the Scheme

The Act says nothing at all about enforcement. Section 108(3) requires the contract to provide that the adjudicator's decision is binding, at least temporarily:

> '108-(3) The contract shall provide that the decision of the adjudicator is binding until the dispute is finally determined by legal proceedings, by arbitration (if the contract provides for arbitration or the parties otherwise agree to arbitration) or by agreement.'

This does not however give any clue to the successful party in the adjudication as to how it is to oblige its opponent to comply with the adjudicator's decision.

The Scheme, which of course only applies if the contract fails to meet the requirements stated in the Act, includes the term that the Act required, and goes further:

> '23-(1) In his decision, the adjudicator may, if he thinks fit, order any of the parties to comply peremptorily with his decision or any part of it.
>
> (2) The decision of the adjudicator shall be binding on the parties, and they shall comply with it until the dispute is finally determined by legal proceedings, by arbitration (if the contract provides for arbitration or the parties otherwise agree to arbitration) or by agreement between the parties.
>
> 24. Section 42 of the Arbitration Act 1996 shall apply to this Scheme subject to the following modifications –
>   (a) in subsection (2) for the word "tribunal" wherever it appears there shall be substituted the word "adjudicator",
>   (b) in subparagraph (b) of subsection (2) for the words

"arbitral proceedings" there shall be substituted the word "adjudication",
(c) subparagraph (c) of subsection 2 shall be deleted, and
(d) subsection (3) shall be deleted.'

It is therefore necessary to turn to the Arbitration Act 1996 to see whether it is of assistance. Modified in accordance with paragraph 24 of the Scheme, section 42 reads thus:

'42. Enforcement of peremptory orders of tribunal

(1) Unless otherwise agreed by the parties, the court may make an order requiring a party to comply with a peremptory order made by the tribunal.

(2) An application for an order under this section may be made –
(a) by the adjudicator (upon notice to the parties),
(b) by a party to the adjudication with the permission of the adjudicator (and upon notice to the other parties), or
(c) [deleted]

(3) [deleted]

(4) No order shall be made under this section unless the court is satisfied that the person to whom the tribunal's order was directed has failed to comply with it within the time prescribed in the order or, if no time was prescribed, within a reasonable time.

(5) The leave of the court is required for any appeal from a decision of the court under this section.'

Presumably the word 'tribunal' in subsection (1) should also have been changed to 'adjudicator', but this was missed.

It is not at all clear how it is intended that this apparent ability to apply to the court for an order requiring compliance is supposed to work in the context of adjudication. If the adjudicator has made a peremptory order, and assuming that 'tribunal' is taken as meaning 'adjudicator', it seems that either the adjudicator or a party (with the permission of the adjudicator) can make an application to the court. But what is a 'peremptory order'? Is the adjudicator's decision a peremptory order?

Peremptory orders in arbitration are introduced by section 41(5) of the Arbitration Act, which states:

'(5) If without showing sufficient cause a party fails to comply with any order or directions of the tribunal, the tribunal may make a peremptory order to the same effect, prescribing such time for compliance with it as the tribunal considers appropriate.'

A peremptory order in arbitration therefore is a second order requiring compliance with an earlier order. The arbitrator may have required one of the parties to produce a particular document or class of documents for inspection, and when the party fails to do so the arbitrator makes an order stated to be peremptory. The reluctant party then knows that if he still declines to comply, an application to the court may well follow.

It is difficult to apply this to an adjudicator's decision. The adjudicator may have decided that X should pay Y £10,000. That is his decision which is delivered to the parties. If X fails to pay, perhaps Y will go back to the adjudicator and ask for a peremptory order. The adjudicator may well respond that he has given his decision and there is nothing further for him to do. He is *functus officio* and has no power to make any further orders.

Some adjudicators make a practice of stating that their decisions requiring payment are peremptory orders and that they give permission for an application to court, so that the successful party can go straight to court to seek an order for compliance. This avoids the problem of seeking a further order from an adjudicator who no longer has any authority, but it is not a peremptory order as is understood under the Arbitration Act.

There is then a further difficulty for the court in deciding on what sort of order to make in order to require compliance with an adjudicator's decision. An injunction requiring the reluctant party to make a payment would be an unusual order for the court to make, but there is no obvious alternative apparent from the legislation or the Scheme.

This difficulty was considered, and an answer found, in the first case that came before the courts involving the Act, *Macob Civil Engineering Ltd* v. *Morrison Construction Ltd* (Sir John Dyson, February 1999). Macob had succeeded in obtaining a decision of an adjudicator that Morrison should pay them £302,366 plus VAT. The adjudication had been carried out under the Scheme, and the adjudicator had expressed his decision as being a peremptory order. Morrison objected on the grounds of breach of natural justice, and also on the basis that there was an agreement to refer disputes to arbitration. These two arguments, considered later in this chapter, were unsuccessful.

Sir John Dyson then considered the problems of dealing with section 42 of the Arbitration Act. He decided that the court could enforce the decision under section 42. There had been argument before him as to whether an injunction should be given, which was the alternative preferred by Macob's counsel, or summary judgment – the preference of Morrison's counsel. Sir John's explanation of the position is helpful guidance as to how this difficult section should be applied, and how applications for enforcement should normally proceed:

> 'I am in no doubt that the court has jurisdiction to grant a mandatory injunction to enforce an adjudicator's decision, but it would rarely be appropriate to grant injunctive relief to enforce an obligation on one contracting party to pay the other. Clearly, different considerations apply where the adjudicator decides that a party should perform some other obligation, e.g. return to site, provide access or inspection facilities, open up work, carry out specified work etc. ... a mandatory injunction to enforce a payment obligation carries with it the potential for contempt proceedings. It is difficult to see why the sanction for failure to pay in accordance with an adjudicator's decision should be more draconian than for failure to honour a money judgment entered by the court.
>
> Thus, section 42 apart, the usual remedy for failure to pay in accordance with an adjudicator's decision will be to issue proceedings claiming the sum due, followed by an application for summary judgment.'

It seems therefore that the peremptory order approach suggested by paragraph 24 of the Scheme will only be followed where the adjudicator's decision requires something other than a payment of money. Even then, the potential difficulty of seeking a further peremptory order from an adjudicator who no longer has authority will have to be addressed. Where, as in most cases, the successful party seeks money, the route to be followed will be the issue of court proceedings and an application for summary judgment under CPR Part 24, considered below.

It is only the Scheme that struggles with complex enforcement provisions. Other standard adjudication rules simply state that the parties may seek summary enforcement, thereby indicating that Sir John Dyson's suggested route is to be preferred.

9.2 Application for summary judgment

Macob Civil Engineering Ltd v. *Morrison Construction Ltd* established that the conventional way to enforce an adjudicator's decision would be through an application for summary judgment. The courts have however gone further and have made it clear that the normal timetable for such an application will be foreshortened dramatically in appropriate circumstances.

The procedure for such an application is set out in Part 24 of the Civil Procedure Rules, which came into force in April 1999. The procedure is similar to that established under Order 14 of the Rules of the Supreme Court which Part 24 superseded. The grounds for summary judgment are set out at rule 24.2:

> '24.2 The court may give summary judgment against a claimant or a defendant on the whole of a claim or on a particular issue if –
> (a) it considers that -
>   (i) that claimant has no real prospect of succeeding on the claim or issue; or
>   (ii) that defendant has no real prospect of successfully defending the claim or issue; and
> (b) there is no other reason why the case or issue should be disposed of at a trial.'

Clearly it is only summary judgment against a defendant that it is likely to be relevant in the context of adjudication.

Rule 24.4 deals with the timetable. A claimant may not apply for summary judgment until the defendant has filed an acknowledgement of service or a defence. The time normally allowed for the defendant to file an acknowledgement, which would normally precede the defence, is 14 days from the effective date of service of the issued claim document. Furthermore, the defendant must be given 14 days' notice of the date fixed for the hearing.

The normal position therefore is that a claim will be issued in court and served on the defendant. The claimant will be unable to take any further step while waiting for the defendant to file an acknowledgement of service, which might take 14 days. If the claimant is able to issue its application for summary judgment immediately thereafter, there will be at least another 14 days before the hearing of the application. The claimant would expect to wait at least 28 days from issuing the claim before appearing in court to apply for summary judgment. This may seem somewhat frustrating for a claimant who has obtained a favourable decision from an adjudicator in possibly rather less time.

The determination of the Technology and Construction Court to support the adjudication process, if there was any doubt following the decision of Sir John Dyson in *Macob Civil Engineering Ltd* v. *Morrison Construction Ltd,* was clearly demonstrated in the second case to be reported, *Outwing Construction Ltd* v. *H. Randell and Son Ltd* (Judge LLoyd, March 1999). Judge LLoyd was not prepared to allow the normal timetable for applications for summary judgment to delay the rapid enforcement of an adjudicator's decision.

Outwing had obtained a decision of an adjudicator on 12 February 1999 that it was entitled to be paid some £16,000. Randell did not pay and said that it wished to challenge the decision. Outwing took no formal action until 8 March 1999, when it issued a writ (this was before CPR displaced the former Rules of the Supreme Court – there is no effective difference to procedures described here).

Two days later, on 10 March, Outwing obtained permission from the court to make an application on very short notice. It issued a summons (now called an application) which was listed for hearing on 12 March. In that application it sought an order that the time for filing an acknowledgement of service should be reduced from 14 days from service of the writ (22 March at the earliest) to just two days after the hearing of the application (i.e. 14 March). That would mean that the application for summary judgment could be issued eight days early, on 14 March. Outwing went further. It also asked for an order that the time for Randell to serve evidence in opposition to the application for summary judgment be restricted to seven days from the date of issue of the application.

Randell decided not to maintain its challenge to the decision, no doubt having had an opportunity to consider the very recent decision in *Macob Civil Engineering Ltd* v. *Morrison Construction Ltd.* On the morning of the hearing of the application to foreshorten the timetable, Randell paid the sum claimed together with interest and the scale costs automatically endorsed on the writ. No doubt Randell thought that the payment would end the matter. Outwing, however, pressed for payment of further costs, covering all the work involved in the application for summary judgment and the efforts to shorten the timetable.

Counsel for Outwing argued that all the steps that had been taken were justified. Randell's counsel raised a formal argument based on the costs demand in the writ but also argued that Outwing had acted with undue haste. There was nothing in the legislation to suggest that the timetable established by the court rules should be changed when an application was made to enforce an adjudicator's decision.

The judge was obliged to consider the formal powers under the court rules to abridge time, and also the formal rules as to costs. His analysis led him to conclude that he had such powers. His consideration of whether or not it was appropriate for him to exercise those powers is of particular interest. Whereas applications to recover ordinary commercial debts are subject to a timetable that allows the defendant to take stock of its position in case there is a defence to the claim, adjudicators' decisions are different:

> 'Thus before a writ is issued (whether or not declared to be a peremptory order) there will normally have been careful consideration of the underlying dispute, its ramifications and of the adjudicator's decision by all parties. The defendant's room for manoeuvre and its need for further time will be limited...
>
> ...there is seemingly no reason why a party who has not voluntarily complied with a decision that it should now honour an outstanding contractual obligation to pay should be allowed the best part of a month, at the very least, before a decision requiring payment to the claimant is converted into the order of a court.'

The claimant succeeded in recovering costs, although a distinction was drawn between costs incurred in preparation of the writ, which were covered by the fixed scale costs, and the proper costs of preparing the application, which were recoverable without reference to a scale.

Judge LLoyd made it clear that the abridgement of time will not automatically be given when application is being made to enforce an adjudicator's decision. There may be circumstances in which a defendant can reasonably require more time than was made available to Randell. The tone of his comments though suggest that it is more likely than not that the court will allow a claimant to proceed very quickly. In the course of the judgment he allowed us a valuable insight into the policy of the Technology and Construction Court:

> 'In anticipation of the need for adjudication matters to be heard without undue delay the judges of this court decided last year that an application marked as concerning adjudication would, if possible, be heard speedily.'

By February 2000, when Judge LLoyd heard another application to enforce an adjudicator's decision (*F.W. Cook Ltd* v. *Shimizu (UK) Ltd*), he felt able to describe the application to abridge time for service as 'usual'.

A claimant wishing to enforce an adjudicator's decision will now normally proceed as follows:

(1) Issue a claim in the High Court (either in London or in a District Registry) marking it as 'Technology and Construction Court Business', and 'In the Matter of the Housing Grants, Construction and Regeneration Act 1996 and in the Matter of an Adjudication'.
(2) At the same time, issue an application:

   (a) to shorten the time limited for filing an acknowledgement of service to two days from the date of hearing of the application,
   (b) to shorten the period of notice of an application for summary judgment from 14 days to say 7 days, and
   (c) to require the defendant to serve any written evidence upon which it intends to rely at the hearing of the application for summary judgment within say five days from service of the notice of the application for summary judgment
   (d) for costs.

   The application should be listed for hearing in 2 or 3 days following its issue.
(3) Immediately serve, preferably by hand delivery, the claim and the notice of application to abridge time. It would be appropriate also to serve in draft the proposed application for summary judgment and the supporting evidence.

There is of course no reason why the claimant should not follow the summary judgment procedure with its normal timetable if it does not wish to deal with the application in this expedited manner.

9.3 Other enforcement procedures

The Technology and Construction Court has enthusiastically supported the use of applications for summary judgment to enforce the decisions of adjudicators, but it may be appropriate in particular circumstances to consider other methods of enforcement.

9.3.1 Injunction

As was made clear in the passage from the judgment of Sir John Dyson in *Macob Civil Engineering Ltd* v. *Morrison Construction Ltd* quoted above, the courts will not normally give an injunction to enforce a decision that a party should make a payment to another party, as the summary judgment procedure is adequate and effective. If however the adjudicator's decision is that some other action is to be taken, such as the opening up of work for tests, there is no reason why a mandatory injunction should not be given.

9.3.2 Winding up petition

There has been no reported use of winding up procedures to enforce payment of adjudicators' decisions, but the procedure can be used. A creditor wishing to petition for the winding up of a debtor company will normally do so on the ground that the debtor is unable to pay its debts (section 122(f) of the Insolvency Act 1986). This can be demonstrated by the service of a statutory demand, pursuant to section 123 of that Act. If the debt remains unpaid after 21 days, the debtor company is deemed by virtue of the statute to be unable to pay. It will not be attractive to use this procedure to pursue the payment of money decided to be due under an adjudicator's decision, but the service of the 21 day notice is not obligatory. This was made clear in the case of *Cornhill Insurance plc* v. *Improvement Services Ltd* (1986), in which Mr Justice Harman said:

> 'where a company was under an undisputed obligation to pay a specific sum and failed to do so, it could be inferred that it was unable to do so; that, accordingly, the defendants could properly swear to their belief in the plaintiff company's insolvency and present a petition for its winding up...'

This approach was followed by the Court of Appeal in *Re Taylor's Industrial Flooring Ltd* (1990). In that case the court went further. Lord Justice Dillon added that the reason for non-payment has to be substantial: 'It is not enough if a thoroughly bad reason is put forward honestly.'

It would seem therefore that if an adjudicator has decided that a sum of money is to be paid by a limited company, and that the time for payment has arrived or has passed, a petition to wind that company up can be presented without any statutory notice being served. Some caution is required however, because if the debtor company does have a 'substantial' reason for non-payment – such as an arguable case that the adjudicator acted in excess of his jurisdiction – the company will resist the petition and may obtain an injunction against the petition proceeding. Substantial costs will be incurred and may be payable by the claimant.

9.3.3 Application to arbitrator

Before the decision in *Macob Civil Engineering Ltd* v. *Morrison Construction Ltd*, it was feared that the courts would be unable to deal effectively with the enforcement of an adjudicator's decision where the underlying contract had contained an agreement to refer disputes to arbitration. This perceived difficulty, and Sir John Dyson's solution, are considered later in this chapter.

If the fears had been well founded, a claimant in such circumstances would have been obliged to commence arbitration proceedings, secure the appointment of an arbitrator (either by agreement or through the appropriate appointing body) and apply to him for an award. It is possible that this route may still be considered appropriate in some cases, particularly if an arbitration has already been commenced and has been running in parallel with the adjudication proceedings, so that there is no delay whilst an arbitrator is appointed.

The provision of the Arbitration Act 1996 that is most likely to be of assistance is section 47:

'47. Awards on different issues, &c.
(1) Unless otherwise agreed by the parties, the tribunal may make more than one award at different times on different aspects of the matters to be determined.

(2) The tribunal may, in particular, make an award relating-
   (a) to an issue affecting the whole claim, or
   (b) to a part only of the claims or cross-claims submitted to it for decision.
(3) If the tribunal does so, it shall specify in its award the issue, or the claim or part of a claim, which is the subject matter of the award.'

Hence if the arbitration is dealing with claims for an extension of time, return of liquidated and ascertained damages, payment of loss and expense and the value of several variations, and the claimant has obtained an adjudicator's decision dealing with one or more aspects of those claims, the claimant may seek an award on his entitlement to immediate payment under that decision. The entitlement under the decision is an aspect of the matters to be determined, and having made an award on that aspect, the arbitrator can continue with final resolution of the claims.

There is no reason why such an application could not be dealt with by the arbitrator rapidly. Indeed the process might be significantly faster than any court procedure, even if expedited as in *Outwing Construction Ltd* v. *H. Randell & Son Ltd.* The whole process might take no more than a few hours.

If there is no current arbitration, it will be necessary for a claimant wishing to use arbitration procedures to commence an arbitration and the reference may be limited to the simple issue of the entitlement under the adjudicator's decision. In that case of course there would be no need to rely on section 47 because the arbitrator's award would be final. Any attempt by the respondent to delay matters by introducing other issues could be defeated by the use of section 47.

Having obtained as award of an arbitrator, either a 'partial award' under section 47 of the Arbitration Act or a final award, the claimant has then to enforce it. This is governed by section 66 of the Arbitration Act:

> '66. Enforcement of the award
>
> (1) An award made by the tribunal pursuant to an arbitration agreement may, by leave of the court, be enforced in the same manner as a judgment or order of the court to the same effect.
>
> (2) Where leave is so given, judgment may be entered in terms of the award.
>
> (3) Leave to enforce an award shall not be given where, or to the extent that, the person against whom it is sought to be enforced shows that the tribunal lacked substantive jurisdiction to make the award.
>
>   The right to raise such an objection may have been lost (see section 73).
>
> (4) Nothing in this section affects the recognition or enforcement of an award under any other enactment or rule of law, in particular under Part II of the Arbitration Act 1950 (enforcement of awards under Geneva Convention) or the provisions of Part III of this Act relating to the recognition and enforcement

of awards under the New York Convention or by an action on the award.'

9.4 Challenges to enforcement

The majority of reported cases dealing with adjudication have been concerned with enforcement of the adjudicators' decisions, and in particular attempts to avoid such requirements (mainly unsuccessful) by those who are required by those decisions to make payments.

9.4.1 Reliance on agreement to arbitrate

During the months immediately prior to the coming into force of the Act, there was considerable concern that decisions would prove to be unenforceable if there was an arbitration agreement in the construction contract. This concern had not been expressed at the time that the Act was passed, but developed as it became clear that the ability of the courts to exercise discretion in deciding whether or not to stay proceedings pending arbitration was going to be much more limited under the Arbitration Act 1996 than had been anticipated.

The position prior to the Arbitration Act 1996 was governed by the Arbitration Act 1950, section 4(1) which provided:

'If any party to an arbitration agreement, or any person claiming through or under him, commences any legal proceedings in any court against any other party to the agreement, or any person claiming through or under him, in respect of any matter agreed to be referred, any party to those legal proceedings may at any time after appearance, and before delivering any pleadings or taking any other step in the proceedings, apply to that court to stay the proceedings, and that court or a judge thereof, if satisfied that there is no reason why the matter should not be referred in accordance with the agreement, and that the applicant was, at the time when proceedings were commenced, and still remains, ready and willing to do all things necessary to the proper conduct of the arbitration, may make an order staying the proceedings.'

A common reason for the court to decline to make such an order was that there was no dispute. This might be established by a successful application for summary judgment under RSC Order 14, the forerunner of Part 24. Hence it was common for defendants to seek a stay of the proceedings under section 4 of the Arbitration Act 1950, and plaintiffs would seek summary judgment. The two applications would be heard together. If the plaintiff was successful in establishing that there could be no defence to the claim, the court would also decide that there was in effect no dispute, and there would be no stay.

Whilst this was common practice in the Official Referees' Courts dealing with UK contracts, the approach of other courts dealing with arbitration with an international element, to which Section 1 of the Arbitration Act 1975 applied, was rather different. Section 1 of the 1975 Act read as follows:

> 'If any party to an Arbitration Agreement to which this section applies ... commences any legal proceedings ... in respect of any matter agreed to be referred, any party to the proceedings may ... before delivering any pleadings or taking any other steps in the proceedings, apply to the Court to stay the proceedings and the Court unless satisfied ... that there is not in fact any dispute between the parties with regard to the matter agreed to be referred, shall make an order staying the proceedings.'

If it were shown that there was an effective arbitration agreement, the only possible basis for declining to stay the action would be that there was no dispute. The court had no discretion in the matter. It was not possible to establish that there was no dispute merely by showing that there was no defence. Mr Justice Saville, later to become Lord Justice Saville and to be responsible for drafting the new Arbitration Act, explained his views in his judgment in *Hayter* v. *Nelson and Home Insurance Co* (1990). His words have become well known:

> 'Two men have an argument over who won the University Boat Race in a particular year. In ordinary language they have a dispute over whether it was Oxford or Cambridge. The fact that it can be easily and immediately demonstrated beyond any doubt that the one is right and the other wrong does not and cannot mean that the dispute did not in fact exist. Because one man can be said to be indisputably right and the other indisputably wrong does not, in my view, entail that there was therefore never any dispute between them.'

In international disputes, therefore, litigation involving a matter subject to an arbitration agreement would almost certainly be stayed. In a domestic matter, such as those normally before the Official Referees' Courts, summary judgment would normally be available.

At first it was not appreciated that the Arbitration Act 1996 was to have any significant change in the way the courts were to approach these applications. Section 9 of the Act provided:

> '(1) A party to an arbitration agreement against whom legal proceedings are brought (whether by way of claim or counter-claim) in respect of a matter which under the agreement is to be referred to arbitration may (upon notice to the other parties to the proceedings) apply to the court in which the proceedings have been brought to stay the proceedings so far as they concern that matter...
>
> (4) On an application under this section the court shall grant a stay unless satisfied that the arbitration agreement is null and void, inoperative, or incapable of being performed.'

This was clearly limiting the discretion that the court had felt free to exercise in UK construction contract matters, but such discretion was apparently protected by section 86 of the Act, which referred to 'domestic' arbitration agreements, involving no parties from outside the UK. It provided:

> '(2) On an application under that section in relation to a domestic arbitration agreement the court shall grant a stay unless satisfied –
> (a) that the arbitration agreement is null and void, inoperative, or incapable of being performed, or
> (b) that there are other sufficient grounds for not requiring the parties to abide by the arbitration agreement.'

This section appeared to allow the Official Referees' Courts to continue exercising a discretion. The fact that the plaintiff was entitled to summary judgment would surely be sufficient to provide 'other sufficient grounds'. It was assumed that despite the changes in specific words in the new legislation, the practice would continue as before. But section 86 was not brought into force with the rest of the Arbitration Act. It had been included in the Act in order to allow further discussion, and those advising the government at the time, led by Lord Saville, whose views had not changed since his decision in *Hayter* v. *Nelson*, recommended bringing domestic arbitration into line with international practice.

It was therefore clear that there might be problems in relying on the summary judgment procedure to enforce an adjudicator's decision if there was an arbitration agreement in the contract. Concern deepened in July 1997 when judgment was given in the Admiralty Court in *Halki Shipping Corporation* v. *Sopex Oils Ltd*. Halki brought a claim under a charterparty for demurrage. Sopex was unable to produce a convincing defence to any but a small part of the claim, but sought an order staying the proceedings under section 9 of the Arbitration Act. It was successful. Effectively a refusal to pay was sufficient to give rise to a dispute and if a party wished to enforce an agreement to refer such disputes to arbitration, it was entitled to such an order.

Many assumed that this was bound to apply to applications to enforce adjudication decisions. A reluctant party would merely have to say that there was a dispute about whether the adjudicator's decision was a valid decision in accordance with the Housing Grants, Construction and Regeneration Act, or in accordance with the relevant rules such as the Scheme, and the courts would not be able to deal with it. There would be a stay of the proceedings under the Arbitration Act.

This point was put to the test in the first case to arise under the Act. In *Macob Civil Engineering Ltd* v. *Morrison Construction Ltd* (Sir John Dyson, February 1999) Macob was a groundworks subcontractor to Morrison. The adjudicator found in favour of Macob who wished to enforce the decision, although not through the mechanism of an application for summary judgment.

Morrison argued that the decision was invalid because, they said, the adjudicator had not followed the rules of natural justice. Morrison gave notice requiring the dispute about the validity of the decision to be referred to arbitration, and when Macob issued proceedings to enforce, Morrison issued a summons to stay the proceedings under section 9 of the Arbitration Act 1996. Counsel for Morrison accepted that a dispute about the issues raised by the adjudication would not be a matter that would prevent enforcement, but if there was a dispute about the validity of the decision itself, that dispute was caught by the agreement to arbitrate.

Sir John Dyson described that argument as 'ingenious', despite the fact that it had been anticipated by most commentators on the new legislation for some two years before the case was heard. Nevertheless he rejected it. He said that Morrison could either have accepted that the decision was valid, albeit wrong, and referred it to arbitration, or alternatively they could say that it was invalid. If they chose that route they would not be able to refer it to arbitration, because it effectively did not exist, but they could try to argue that it should not be enforced. They could not do both. By referring it to arbitration they had accepted that it was a valid decision and could no longer argue the opposite.

Those who had been described by Sir John as 'ingenious' may have felt able to return the compliment. He had surmounted the difficulty that they had felt insurmountable. On the basis of this decision, it is not possible to argue that an adjudicator's decision is not enforceable by the court if there is an arbitration agreement between the parties. The defendant in *Absolute Rentals Ltd* v. *Gencor Enterprises Ltd* (July 2000) failed to persuade Judge Wilcox to apply *Halki* v. *Sopex*. It may be that the Court of Appeal will, if asked, have some difficulty with Sir John Dyson's analysis, but until a case involving this question is taken to appeal, parties are likely to be advised that an objection to enforcement based on an arbitration agreement will fail.

If the decision in *Macob* is effectively overturned in a later case, all is not necessarily lost for the party wishing to enforce a decision in such circumstances. If the contract is on a JCT form or one if its associated subcontracts, the enforcement of an adjudicator's decision is outside the scope of the arbitration agreement and therefore would not be subject to an application for a stay under the Arbitration Act. The ICE and CECA contract and subcontracts make a similar exclusion.

Even if the party with the benefit of an adjudicator's decision is obliged to go to arbitration to seek enforcement of that decision, there may not be too serious a delay. Sir John Dyson, in *Macob*, was sceptical about this. He said:

> 'I accept that arbitration can be swift but often it is not, and, as already explained, in some cases it cannot even be started until long after the dispute has arisen. More fundamentally, if Parliament had thought that resolution by arbitration was a swift and effective procedure, it would surely not have seen the need to enact the Act at all.'

Nevertheless, section 47 of the Arbitration Act makes it clear that unless the power is excluded by agreement, an arbitrator may make awards on different issues in the reference at different times. If therefore a party commences an arbitration disputing the validity of an adjudicator's decision as well as raising the fundamental matters in dispute, perhaps with a view to delaying the enforcement of the decision until the whole dispute is revisited, the arbitrator can resolve the enforceability of the decision swiftly, moving on to the original issues in due course. An arbitrator should be able to act as quickly as the Technology and Construction Court in such matters, and provide a service as effectively as the courts acting under Part 24.

9.4.2 Error in the decision

The fact that the adjudicator made an obvious error in his decision does not render the decision unenforceable.

In *Bouygues UK Ltd* v. *Dahl-Jensen UK Ltd* (Sir John Dyson, November 1999), Dahl-Jensen had made a claim in an adjudication for approximately £2.9 million for breaches of contract relating to late provision of information, £2.1 million for additional works and £225,000 for other breaches of the subcontract. Bouygues counter-claimed approximately £1.16 million for overpayments, £315,000 liquidated damages and £3.9 million other damages. The adjudicator found that both parties succeeded in parts of their claims, and that a net sum of £207,741 was due to Dahl-Jensen.

The problem arose from the way in which the figures had been calculated. The adjudicator calculated the total of the sums due to Dahl-Jensen without making any deduction for retention. He then deducted the sums that had been paid as interim payments. Those sums of course were net of retention. The effect of this calculation was to release retention, which was not yet due for release. If retention had been deducted from the money that he had calculated to be due to Dahl-Jensen, the effect would have been to change the apparent underpayment to an overpayment. Dahl-Jensen would have been obliged to pay Bouygues £141,254.

Bouygues argued that by effectively releasing retention, the adjudicator had gone outside his jurisdiction. He had not been asked to release retention, but that is what he had done. Because he had done something that he was not appointed to do, his decision should not be enforced.

This argument failed, not because it was wrong in theory but because the court concluded that the adjudicator had not in fact come to his decision on the basis that he was releasing retention. There was nothing in the reasons that accompanied the decision or in the other material before the court to suggest that he had had any intention to do that. The court decided that he had made a simple mistake in the way that he exercised the jurisdiction that he indisputably had. Sir John Dyson approached the question robustly:

'...in deciding whether the adjudicator has decided the wrong question rather than given a wrong answer to the right question, the court should bear in mind that the speedy nature of the adjudication process means that mistakes will inevitably occur, and, in my view, it should guard against characterising a mistaken answer to an issue that lies within the scope of the reference as an excess of jurisdiction.'

The test that was applied was to ask whether the adjudicator had answered the right question in the wrong way, or the wrong question altogether. If it was the former, he had acted within his jurisdiction and the decision was enforceable. If the latter, he had acted outside his jurisdiction and the decision was not enforceable. This was the test that had been applied in *Nikko Hotels (UK) Ltd* v. *MERPC Plc* (1991), a case that concerned the decision of an expert valuer.

The *Bouygues* case went on to the Court of Appeal, which entirely supported the approach taken by Sir John Dyson.

It was assumed that the adjudicator in *Bouygues* had made an arithmetical error but there is nothing in the judgments of the Technology and Construction Court to support that assumption. The adjudicator may have made an error, or alternatively he can simply have made his calculations without including retention because he considered it appropriate to do so after a determination of the contract. Such a calculation would not have been a conscious decision to release retention but merely an arguably misguided approach to the quantification of the claim. All that the court knew on the basis of the material before it was that the adjudicator did not think that he had made a clerical error. He had been invited to correct his decision on the basis that it contained such an error and he had declined to do so.

In *Tim Butler Contractors Ltd* v. *Merewood Homes Ltd* (April 2000) Judge Gilliland was asked to enforce the decision of an adjudicator who had decided that there was a right to interim payments. There had been a programme that showed that work would take less than 45 days, but the adjudicator had decided that there was no evidence that this programme had been agreed, or that there was a term of the contract that the work would be completed in a shorter time. The defendant argued that the adjudicator had been clearly wrong, and that therefore he had acted outside his jurisdiction and the decision should not be enforced. The judge rejected the argument, without saying whether he thought that the adjudicator was right or wrong. The matter was within his jurisdiction to decide and therefore the decision was enforceable.

Those who wish to argue that in making a mistake an adjudicator effectively goes outside his jurisdiction may take some comfort from the words of Lord Kingarth in *Allied London and Scottish Properties plc* v. *Riverbrae Construction Ltd* (July 1999). The case concerned the failure of the adjudicator to order that the money he had found to be due to the claimant should be paid into a secure stakeholder account pending the resolution of other related claims. The judge found that there was no reason why the adjudicator should have made such an order, but went on to say:

> 'Whatever wide powers may be given to adjudicators to facilitate speedy resolution of the disputes before them, no power is given to make decisions contrary to the rights or obligations of the parties arising as a matter of law.'

In another Scottish case, *Homer Burgess Ltd* v. *Chirex (Annan) Ltd* (November 1999, Lord Macfadyan, Court of Session), it was held that an adjudicator had made a mistake in his decision that pipework connecting various items of machinery and equipment on a site where the primary activity was the processing and production of pharmaceuticals was not 'plant'. If it was plant it was excluded from the ambit of the Act (see Chapter 2). The judge decided that the pipework was plant, and that the 'disputes relating to that work were therefore not disputes on which the adjudicator had power to make a decision.'

Judge Gilliland dealt with the *Homer Burgess* case in some detail in *Tim Butler Contractors Ltd* v. *Merewood Homes Ltd*. He accepted that an adjudicator could not make a binding decision on whether or not a contract was a construction contract and therefore whether or not the contract was subject to the adjudication process, but the dispute in *Tim Butler Contractors* did not concern questions of jurisdiction.

It may be that the Scottish courts will take a more restrictive approach to the enforcement of adjudicators' decisions that are seen to be obviously mistaken, but in England and Wales a mistake of fact or law is unlikely to be enough to render the decision unenforceable.

9.4.3 Lack of jurisdiction

The courts have made it clear that if the adjudicator did not have jurisdiction to decide the dispute, the decision will not be enforced. Sir John Dyson, in *The Project Consultancy Group* v. *The Trustees of the Gray Trust* (July 1999) said this:

'I conclude, therefore, that is open to a defendant in enforcement proceedings to challenge the decision of an adjudicator on the grounds that he was not empowered by the Act to make the decision.'

Whilst he later made it clear in *Bouygues*, and in particular the passage from that case set out in section 9.2 above, that the court would not artificially describe arguments based on the factual content of decisions as arguments on jurisdiction, lack of jurisdiction is likely to be the most successful argument that can be used by a defendant seeking to avoid enforcement

Jurisdiction may be found wanting in various ways, as follows.

Absence of a contract in writing, as defined by section 107 of the Act

This was the basis of the court's decision in *Grovedeck Ltd* v. *Capital Demolition Ltd* (Judge Bowsher, February 2000), discussed in Chapter 2.

Contract not for construction operations

This was the issue raised in *Palmers Ltd* v. *ABB Power Construction Ltd* (Judge Thornton, August 1999). On the facts of that case, also discussed in Chapter 2, the contract was for construction operations and the adjudicator therefore had jurisdiction. This was however a matter which was appropriate for the court to consider. The case came before the court not as an application to enforce, but an application made during the course of the adjudication for a declaration that the contract was for construction operations.

A mistaken decision that a contract is for construction operations when in fact it is not was the subject of *Homer Burgess Ltd* v. *Chirex (Annan) Ltd* (November 1999, Lord Macfadyan, Court of Session – see above and Chapter 2). Enforcement was refused on the basis that the adjudicator had had no jurisdiction.

A decision based on a compromise agreement was not supported by a Part 24 judgment because the compromise agreement was not itself a construction contract, despite the fact that the original contract, which had given rise to the compromised dispute, was for construction operations (*Lathom Contraction Ltd* v. *Cross and Cross* (October 1999, Judge Mackay) and *Shepherd Construction Ltd* v. *Mecright Ltd* (July 2000, Judge LLoyd) see Chapter 3).

Contract predates 1 May 1998

The Act came into force on 1 May 1998, and any contract formed before that date is not affected by it. If such a contract does not contain adjudication provisions, and therefore relies on the Act and the Scheme to provide a basis for adjudication, there will be no possibility of effective adjudication without express agreement.

The point is perhaps self-evident but formed the basis for the leading case on jurisdiction, *The Project Consultancy Group* v. *The Trustees of the Gray Trust*, mentioned above and discussed in Chapter 2. As both that case and *Christiani & Nielsen Ltd* v. *The Lowry Centre Development Company Ltd* (also discussed in Chapter 2) show, there can be serious doubt about whether the contract was formed before or after the critical date. With the passage of time of course the likelihood of this point arising diminishes.

Dispute not covered by adjudication agreement

It was argued in *Northern Developments (Cumbria) Ltd* v. *J. & J. Nichol* (January 2000) that a repudiation of the contract, accepted by the innocent party, had brought the contract to an end, and that therefore there was no effective agreement to refer the dispute to the adjudicator. This was rejected on principles derived from arbitration (see Chapter 3).

A similar argument had been adopted by the defendant in *A. & D. Maintenance and Construction Ltd* v. *Pagehurst Construction Services Ltd* (June 1999). Judge Wilcox found that the adjudication provisions survived the determination of the subcontract.

Decision outside the terms of reference in the notice of adjudication

This problem was discussed in Chapter 4. If the adjudicator has done something that he was not asked to do, in that he has decided a dispute that was not included in the notice of adjudication, he will have exceeded his jurisdiction and his decision will not be enforced. Although such a situation was found not to have occurred in *F.W. Cook Ltd* v. *Shimizu (UK) Ltd* (February 2000) Judge Humphrey LLoyd made that clear.

The adjudicator should have resigned under paragraph 9(2) of the Scheme

In *Sherwood & Casson Ltd* v. *Mackenzie Engineering Ltd* (November 1999) it was argued that an adjudication on a final account covered essentially the same grounds as a previous adjudication on an interim account, and that therefore the adjudicator should have resigned. This requirement is found in paragraph 9(2) of the Scheme. Accordingly it was said that he had no jurisdiction. Judge Thornton rejected that argument, finding that the final account process is different from the interim valuation procedure and that the disputes were therefore not the same. This case is also considered in Chapter 4.

9.4.4 Concurrent court proceedings

In *Herschel Engineering Ltd* v. *Breen Property Ltd* (Mr Justice Dyson, April 2000) it was argued that if there were concurrent court proceedings dealing with the same issues as the adjudication, an adjudicator's decision should not be enforced pending the outcome of those proceedings. It was also argued that the issue of court proceedings was an effective waiver or repudiation of the right to adjudicate, and that therefore adjudication proceedings could not even be started. Both arguments failed, although the judge was prepared to accept that in some cases, particularly if final judgment was expected shortly, a stay of execution might be given.

9.4.5 Inability to repay or insolvency

In rejecting Breen's arguments in the *Herschel Engineering* case, Sir John Dyson had suggested that a stay might be appropriate if there was evidence before the court that the recipient of the payment would be unable to repay the sum paid if the adjudicator's decision was reversed by subsequent litigation. Breen made a further application on those grounds, heard before Judge LLoyd in July 2000 (unreported). Some evidence was produced of the current financial position of Herschel Engineering, but there was no evidence available of the likely financial position of the company in the future, when an order for repayment might be made. Judge LLoyd found that to be inadequate. Before giving a stay, the court would have to be satisfied that there was a real risk that the claimant would be unable to repay the relevant sum at the time that it was ordered to do so, and evidence as to current financial situation was not sufficient by itself.

A similar approach was adopted by Judge Wilcox in *Absolute Rentals Ltd* v. *Gencor Enterprises Ltd*. He said:

'I am not in a position to judge the financial standing of either company. It is not desirable that I should on such limited evidence... [Adjudication] is a robust and summary procedure and there may be casualties although the determinations are provisional and not final.'

Judge Havery was equally unreceptive to the argument that the recipient of money following the enforcement of an adjudication decision would be likely to dissipate it before final resolution of the dispute and perhaps an order that it be repaid (*Elanay Contracts Ltd v. The Vestry* August 2000).

Mere doubt about the likely financial position of the claimant therefore is not enough. If the recipient company is in formal insolvency procedures, there is of course rather more than mere doubt about its financial position. Any decision made in favour of such a claimant, and any summary judgment of the court to enforce that decision, will be subject to the Insolvency Rules 1986.

This was the unexpected conclusion of the Court of Appeal in *Bouygues UK Ltd v. Dahl-Jensen UK Ltd* (July 2000). Bouygues sought to avoid the consequences of an adjudicator's decision. It had unsuccessfully opposed summary judgment in the Technology and Construction Court, basing its arguments on jurisdiction. It did not raise the question of the Insolvency Rules either in the lower court or before the Court of Appeal. During the course of argument, Lord Justice Chadwick asked counsel to consider the effect of rule 4.90, which reads:

'4.90 Mutual credit and set-off
(a) This Rule applies where, before the company goes into liquidation there have been mutual credits, mutual debts or other mutual dealings between the company and any creditor of the company proving or claiming to prove for a debt in the liquidation.
(b) An account shall be taken of what is due from each party to the other in respect of the mutual dealings, and the sums due from one party shall be set off against the sums from the other.
(c) Sums due from the company to another party shall not be included in the account taken under paragraph (2) if that other party had notice at the time they became due that a meeting of creditors had been summoned under section 98 or (as the case may be) a petition for the winding up of the company was pending.

(d) Only the balance (if any) of the account is provable in the liquidation. Alternatively (as the case may be) the amount shall be paid to the liquidator as part of the assets.'

The effect of this rule, or the comparable provisions of section 323 of the Insolvency Act 1986, had been explained by Lord Hoffman in the House of Lords decision *Stein* v. *Blake* (1996). In that case the plaintiff had sued for damages for breach of contract. The defendant counterclaimed for damages for misrepresentation. The plaintiff was adjudged bankrupt, and the cause of action was assigned with a view to the action being prosecuted outside the bankruptcy. The assignment was declared invalid, because once the bankruptcy had commenced the rights of action for claim and counterclaimed were merged, and there was no cause of action to assign at all until the account had been taken in the bankruptcy proceedings.

Lord Justice Chadwick explained the relevance in *Bouygues*:

'The importance of the rule is illustrated by the circumstances in the present case. If Bouygues is obliged to pay to Dahl-Jensen the amount awarded by the adjudicator, those monies, when received by the liquidator of Dahl-Jensen, will form part of the fund applicable for distribution amongst Dahl-Jensen's creditors. If Bouygues itself has a claim under the construction contract, as it currently asserts, and is required to prove for that claim in the liquidation of Dahl-Jensen, it will receive only a dividend pro rata to the amount of its claim. It will be deprived of the benefit of treating Dahl-Jensen's claim under the adjudicator's determination as security for its own cross-claim.

... bankruptcy set-off requires an account to be taken of liabilities which at the time of the bankruptcy may be due but not yet payable, or which may be unascertained in amount or subject to contingency. Nevertheless the insolvency code requires that the account shall be deemed to have been taken, and the sums due from one party shall be set off against the other, as at the date of insolvency order. Lord Hoffman pointed out also that it was an incident of the rule that claims and cross-claims merge and are extinguished; so that, as between the insolvent and the other party, there is only a single claim – represented by the balance of the account between them. In those circumstances it is difficult to see how a summary judgment can be of any advantage to either party where, as the 1966 Act and paragraph 31 of the [Construction Industry Council] Model Adjudication Procedure make clear, the account can be reopened at some stage; and has to be reopened in the insolvency of Dahl-Jensen.'

The Court of Appeal dismissed Bouygues' appeal against summary judgment because the insolvency point had not been put to either the Technology and Construction Court or to the Court of Appeal, and on the question put by Bouygues in its appeal, Dahl-Jensen had succeeded – the adjudicator's decision had been within his jurisdiction. Nevertheless Bouygues were prevented from enforcing the judgment that they had obtained. The Court of Appeal imposed a stay of execution to allow the insolvency procedures to take their course.

In the light of the comments of Lord Justice Chadwick it must be considered highly unlikely that summary judgment will be given to enforce an adjudication decision where the claimant is in liquidation (or bankruptcy) and the defendant can demonstrate that it has a real prospect of defending the substantive claim. Whether or not summary judgment is given, the claimant in liquidation will not be able to enforce payment.

9.4.6 Set-off (other than in insolvency)

The comments of the Court of Appeal in *Bouygues* about insolvency set-off should not be allowed to confuse consideration of legal set-off when no insolvency is involved. The two are quite different, as was explained by Lord Hoffman in *Stein* v. *Blake*:

> 'Legal set-off does not affect the substantive rights of the parties against each other, at any rate until both causes of action have been merged in a judgment of the court. It addresses questions of procedure and cash-flow...
>
> Bankruptcy set-off, on the other hand, affects the substantive rights of the parties by enabling the bankrupt's creditor to use his indebtedness to the bankrupt as a form of security. Instead of having to prove with other creditors for the whole of his debt in bankruptcy, he can set off pound for pound what he owes the bankrupt and prove for or pay only the balance...'

It is not unusual for a defendant to raise arguments about a right to set off a cross-claim in order to defeat an application for summary judgment. It is well established, through cases such as *Hanak* v. *Green* (1958) and *Modern Engineering (Bristol) Ltd* v. *Gilbert Ash*

(Northern) Ltd (1974), that there are classes of cross-claim which may be used to establish a right of set-off so as to operate as a defence. In the absence of any contractual stipulation, the question that the court will ask is whether there is a close connection with the claim itself so that it would be clearly unjust to allow the claimant to recover on the claim without taking the cross-claim into account. In *Hanak* v. *Green* the claimant sought damages against a builder for failure to complete the contract works. The builder was allowed to set off claims for extra work carried out, loss caused by the claimant's refusal to allow the builder's labour onto site and other damages claimed on the basis of trespass to tools. An example of a cross-claim that would be considered too remote is the claim by a contractor for damages arising on another site under another contract.

An adjudicator's decision typically establishes the right of one party to be paid by the other, at least until the matter is subjected to further examination in legal proceedings or arbitration. It is that right that the successful party will seek to enforce through an application to court under Part 24. There is nothing in the Act to suggest that the right is a special right against which a cross-claim cannot be set off.

This does not mean however that the right of set-off outlined above will prevent an order for enforcement. This may depend on provisions other than those dealing with adjudication.

Part II of the Act deals not only with adjudication but also with payment. Section 111 deals with a party's ability to withhold payment:

'111–(1) A party to a construction contract may not withhold payment after the final date for payment of a sum due under the contract unless he has given an effective notice of intention to withhold payment . . .

(2) To be effective such a notice must specify –
  (a) the amount proposed to be withheld and the ground for withholding payment, or
  (b) if there is more than one ground, each ground and the amount attributable to it,
and must be given not later than the prescribed period before the final date for payment.

(3) The parties are free to agree what that prescribed period is to be.
In the absence of such agreement, the period shall be that

provided by the Scheme for Construction Contracts. [Paragraph 10 of the Scheme provides that the period is 7 days]

(4) Where an effective notice of intention to withhold payment is given, but on the matter being referred to adjudication it is decided that the whole or part of the amount should be paid, the decision shall be construed as requiring payment not later than –
  (a) seven days from the date of the decision, or
  (b) the date which apart from the notice would have been the final date for payment,
whichever is the later.'

In *VHE Construction PLC v. RBSTB Trust Co Ltd* (Judge Hicks, January 2000), the court was asked to enforce decisions of two adjudicators dealing with claims for payment by the contractor. One had had the effect of declaring that the contractor's application for payment under a JCT 81 contract had become due because of the failure of the employer to serve a notice specifying the sum that it proposed to pay within five days of receipt of the application. The second adjudication dealt with the correct computation of the interim payment. The effect of the two decisions was that a reduced sum of money was to be paid on 4 November 1999. The defendant sought to set off a claim for liquidated damages. It contended, unsuccessfully, that various documents had been notices under section 111, and then further argued that it had a residual right to set off liquidated damages. The arguments failed. Judge Hicks said:

'...section 111 now constitutes a comprehensive code governing the right to set off against payments contractually due. RBSTB has not complied with it. It would make a nonsense of the overall purpose of Part II of the Act, to which sections 108 and 111 are central and in which they are closely associated, not least by the terms of section 111(4), if payments required to comply with adjudication decisions were more vulnerable to attack in this way than those simply falling due under the ordinary contractual machinery.'

The timing of the two decisions enabled the second decision to affect the first, and it might therefore be said that there are circumstances in which one decision can be set off against another decision. This was clarified by Judge Bowsher in *Northern Developments Ltd v. J. & J. Nichol* (January 2000, 11 days later than *VHE*):

'If there are two conflicting adjudication decisions, it may be appropriate to set one off against the other in enforcement proceedings, but that is not an authority for making a set-off within adjudication proceedings of matters sought to be introduced in breach of the statutory provisions.'

These decisions, and section 111 on which they are based, deal with claims made by contractors for payment. Clearly such payments are strictly bound by the requirements of the Act, and set-off in the absence of appropriate notices will not be possible. But adjudication is not only concerned with claims for payment. A claim by an employer against an architect for damages for breach of the contractual duty of care under the architect's retainer might be brought in adjudication, and a decision obtained that the architect should pay a sum of money. There would seem to be nothing preventing the architect from raising a claim for professional fees which would form the basis of an argument that the architect was able to claim a set-off, potentially thereby preventing the court from giving summary judgment under Part 24. A decision of an adjudicator that the contractor should pay sums by way of liquidated and ascertained damages might be met by a cross-claim which would have a similar effect.

9.4.7 Failure to act in accordance with natural justice

As discussed in Chapter 6, the attempt in *Macob Civil Engineering Ltd* v. *Morrison Construction Ltd* to oppose an application to enforce an adjudicator's decision on the basis of a failure to observe the rules of natural justice failed. In *A. Straume (UK) Ltd* v. *Bradlor Developments Ltd* (1999) the Chancery Court judge grappled with the nature of the adjudication process and concluded:

'that it is, in effect, a form of arbitration, albeit the arbitrator has a discretion as to the procedure that he uses, albeit that the full rules of natural justice do not apply.'

Neither decision however went so far as to suggest that the rules of natural justice should be considered entirely irrelevant to the adjudication process. In August 2000 Judge Bowsher applied them when declining to enforce an adjudicator's decision (*Discain Project Services Ltd* v. *Opecprime Developments Ltd*). One of the issues in the adjudication had been the validity of a withholding notice under section 111 of the Act. It had been written on the letterhead of an associated company of Opecprime, and Discain said that therefore the notice was ineffective. Discain's representative had telephoned the adjudicator three times and discussed this issue with him. Discain had sent faxes summarising the conversations, and copied those faxes to Opecprime, but one of the faxes had been delayed for three days.

The judge distinguished between formal breaches of natural justice, which clearly did not affect the outcome of the case (such as those considered by Sir John Dyson in *Macob*), and more significant breaches that may have had an effect. On the facts of this case, the judge felt that the private telephone conversations fell into the latter category.

9.4.8 Human Rights Act 1998

The Human Rights Act 1998 came into force on 2 October 2000. Its effect is to incorporate into the law of England and Wales the provisions of the European Convention on Human Rights. The Convention has been with us for a long time. It was prepared as an attempt to avoid a recurrence of the events that led to World War II, and came into force in 1952. Since 1968 individuals from the UK have been able to take their cases direct to the European Court. Now the rights given by the Convention can be enforced in the domestic courts.

The most relevant provision of the Convention is article 6:

'In the determination of his civil rights and obligations or of any criminal charge against him, everyone is entitled to a fair and public hearing within a reasonable time by an independent and impartial tribunal established by law...'

Section 6 of the Human Rights Act refers to public authorities, which are defined in section 6(3):

'In this section "public authority" includes –
(a) a court or tribunal, and
(b) any person certain of whose functions are functions of a public nature...'

Section 6(1) states that it is unlawful for a public authority to act in a way which is incompatible with a Convention right.

It has been suggested that an adjudicator is a public authority within the Act's definition and that therefore it would be unlawful for an adjudicator to act in a way which is incompatible with article 6. If the adjudicator acts unlawfully the court will be unable to give judgment on the basis of the adjudicator's decision. This analysis would require the adjudicator to be very cautious in exercising the initiative that is given to him by the Act, the Scheme and all the various sets of adjudication, and would be likely to lead to a minute examination of the procedures adopted in any application to enforce a decision.

The first question that must be asked is whether or not the adjudicator is indeed a public authority. If the construction contract does not contain adjudication provisions that are compatible with the Housing Grants, Construction and Regeneration Act, the Act will impose the Scheme for Construction Contracts. Whilst the Scheme operates through the mechanism of the contract, it might be argued that the adjudicator is performing a statutory function which is 'of a public nature'. It is less easy to see how that might be argued in the case of an adjudicator who has been appointed by the operation of a purely private contractual procedure. The debate is in some measure similar to that relating to the question of whether adjudicators' decisions can be subject to the process of judicial review, discussed in Chapter 1.

It is unlikely that it will be established that all adjudicators are 'public authorities', but it is possible that in the circumstances of a specific case the adjudicator will be found to be acting as one. If so, the next question arises under article 6. Is the process of producing a decision in adjudication a determination of a person's civil rights or obligations?

Section 108 of the Housing Grants, Construction and Regeneration Act provides that the adjudicator's decision is to be binding 'until the dispute is finally determined' by litigation, arbitration or agreement. With respect to the draftsman of the Act, the expression 'finally determined' is tautologous in that 'determination' is by definition final and conclusive. The adjudicator is therefore not determining anything – he is merely making a decision that the parties will accept until final resolution by a full process. Either party still has the right to go further.

As discussed in Chapter 1, it is suggested that the adjudicator's decision has more in common with the decision of the engineer under clause 66 of the ICE Standard Form, than with a trial or arbitration. If that analysis is correct, the Human Rights Act does not apply to the adjudication process. At the time of writing

(November 2000) there has been no reported case directly on the point, but in *Elanay Contracts Ltd* v. *The Vestry* (August 2000, Judge Havery) the defendant relied on the Convention, the Human Rights Act not yet being in force. It argued that the period in which the adjudicator had been obliged to come to his decision was so short that its right under article 6 of the Convention to a reasonable opportunity to present its case had been infringed. There were particular problems for The Vestry. It had been unable to deal with the preparation of its case because of the unavailability of one of its employees due to personal circumstances. Moreover Elanay had delivered documents late but had only agreed a very limited extension of time for the adjudication. The judge held that article 6 did not apply to adjudication because the process was not public and did not involve a final determination of the rights of the parties. There is no reason to believe that a defence based on the Human Rights Act, as opposed to the Convention, would not suffer a similar fate.

CHAPTER TEN
PAYMENT

10.1 Introduction

Whereas the provisions of the Act regarding adjudication were perhaps the most innovative, introducing a completely new method of dispute resolution for the construction industry quite unlike any procedures that had previously been available, the provisions regarding payment were at least as controversial. The Act sought to impose terms for payment on commercial contracts regardless of whether or not the parties were of equal or unequal bargaining power. This was not an attempt to protect the small business from the large who might have been abusing its commercial dominance by insisting on unreasonably favourable payment terms. The relative size of the parties is completely irrelevant. There will be many occasions where the benefit of the Act is given to the large subcontractor in dealing with the small main contractor, or the very large construction company carrying out work for the relatively small client.

The provisions of the Act with regard to payment apply in the same circumstances as the provisions regarding adjudication. It is therefore necessary to consider whether the relevant contract is for 'construction operations' and whether it is a 'construction contract', terms defined in sections 104 and 105 of the Act. The payment provisions do not apply to contracts excluded from the operation of Part II of the Act (section 106), and only apply to contracts that are agreements in writing, section 107. These matters were discussed in Chapter 2.

The payment provisions differ from those relating to adjudication in the way in which they are incorporated into the contract. The contract is required to comply with the requirements of the Act regarding adjudication. If it does not comply with all the requirements, the adjudication provisions of the Scheme for Construction Contracts apply to the contract. The contract may contain an adjudication procedure which complies with some but not all of the Act's requirements. Nevertheless the Scheme's adjudication procedures will be imposed and will replace any contractual provision that is in conflict with the Scheme.

This is not the case with the provisions as to payment. Any contractual provision which complies with the Act will remain effective, and the Scheme will only operate to replace non-compliant provisions or to fill a gap where no provision is made at all.

There are also some provisions of the Act relating to withholding payment, suspension of performance, and conditional payment provisions, which are mandatory regardless of the terms of the contract. They do not rely for their application on the implication of terms into the contract.

10.2 The right to stage payments

'109-(1) A party to a construction contract is entitled to payment by instalments, stage payments or other periodic payments for any work under the contract unless –
   (a) it is specified in the contract that the duration of the work is to be less than 45 days, or
   (b) it is agreed between the parties that the duration of the work is estimated to be less than 45 days.

(2) The parties are free to agree the amounts of the payments and the intervals at which, or circumstances in which, they become due.

(3) In the absence of such agreement, the relevant provisions of the Scheme for Construction Contracts apply.

(4) References in the following sections to a payment under the contract include a payment by virtue of this section.'

Most forms of contract in the construction industry, whether standard forms or custom drafted, include provisions for payments on account as the work progresses, and many will have assumed that where nothing specific is stated, any contract for work lasting longer than say a couple of months would have an implied entitlement to periodic payments. It was by no means established however that this was the case, and at best it could be said that in some circumstances a term might be implied to that effect. Section 109 has therefore made the position much more clear.

The right to some form of periodic payment is absolute unless the contract specifies that the duration of the work is to be less than 45 days or the parties agree that it is estimated to be less than 45 days. Clearly therefore if it is a term of the contract that the work will be complete in a shorter period, there will be no right to an interim payment. The agreement to an estimated time of less than 45 days may be much less clear. The agreement need not necessarily be contained in the contract itself. It is even possible that the contract period, stated in the contract, will be for a longer period, but the parties may agree orally that notwithstanding the obligation to complete in such longer period, it is likely that the works will be complete more quickly. This agreement might be made before or after works have started, although it is perhaps unlikely that a party will voluntarily give up a right to interim payment after work has started by agreeing that the duration is likely to be less than 45 days.

The meaning of 'duration' is not entirely clear. If work is to be carried out in two site visits, each lasting ten days, over a period of two months, there may be an argument that the duration of the work is in fact 20 days. Alternatively it may be said that the duration is 60 days. It is submitted, but without any authority, that the latter is probably the correct interpretation.

It is also unclear what the difference is between 'instalments, stage payments or other periodic payments'. It is possible that the three terms, which might all mean the same thing, were included in order to make it clear that any contractual provision is acceptable, so long as it involves payment on some basis other than one payment at the end of the job. In this, and in what follows in subsection 110(2), the maximum flexibility is allowed to contracting parties to make arrangements that suit their circumstances.

If the 45 day exemption does not apply, the parties are free to agree how much is to be paid and when. The most onerous payment terms might be included and still satisfy the Act's requirements. For example, in a contract for work to a value of £1 million to be carried out over six months, it might be agreed that there will be one payment of £5 after five months, and the balance paid at the end (or substantially later than the end) of the job. Such terms have not been common before the Act and there is no reason why they should become popular now, but they are perfectly compatible with the Act. The important point is that the Act allows maximum flexibility in payment arrangements. If however the question of interim payments is not addressed at the time of the contract, the Scheme will apply and will impose its regime.

The Scheme deals with the timing of periodic payments and the quantification of them together. The Scheme's provisions are considered below.

10.3 Timing and quantification of payments

> 'S 110–(1) Every construction contract shall –
> (a) provide an adequate mechanism for determining what payments become due under the contract, and when, and
> (b) provide for a final date for payment in relation to any sum that becomes due
>
> The parties are free to agree how long the period is to be between the date on which a sum becomes due and the final date for payment.
>
> (2)...
>
> (3) If or to the extent that a contract does not contain such provision as is mentioned in subsection (1) or (2), the relevant provisions of the Scheme for Construction Contracts apply.'

Once again the Act allows maximum flexibility in enabling the parties to make such arrangements for payment as they think appropriate or can negotiate. Any 'mechanism' can be stipulated – the only requirement is that it should be 'effective'. The parties must be able to work out how much is due and when it is due. Payments may be calculated on a straight line basis, on the basis of the value of work carried out, stages reached in the construction process, or on any other basis, so long as the terms are agreed at the time that the contract is made.

There is, as yet, no judicial comment on what does or does not constitute an adequate mechanism, perhaps because an adjudicator's decision on the point is unlikely to be susceptible to challenge in summary proceedings. It is suggested that the words should be given their normal common sense meaning. A contract can certainly have a mechanism for such matters which is inadequate. One example can be found in the facts that lay behind the first reported decision involving the Act, *Macob Civil Engineering Ltd* v. *Morrison Construction Ltd* (Mr Justice Dyson, February 1999). The subcontract between the parties had contained conflicting terms as to the dates for payment in the several subcontract documents. Having heard evidence on the point, the adjudicator had decided that he could not establish what the terms for payment were, if indeed any had been agreed at all. He concluded that there was therefore no adequate mechanism in place, and that the Scheme's provisions should apply.

Subcontractors sometimes argue that a clause in their subcontracts linking payment of retention to certification under the main contract, which may of course be delayed by circumstances for which the subcontractor has no responsibility whatever, is not an adequate mechanism for establishing when payment of retention is to be made. This will be considered further when dealing with conditional payments or 'pay-when-paid' clauses in section 10.7.

Time for payment must be dealt with in two stages. The first, under subsection 110(1)(a), is the time at which a payment becomes due. This is the date at which the entitlement to payment crystallises, but it is not the date at which it is to paid. That involves a second stage, required under subsection 110(1)(b). At some stage after the payment becomes due, the final date for payment will be reached. This is the end of the 'credit period' and payment must be made no later than that date. Each sum due, whether an interim or final payment, will have a final date for payment. The concept is important. The final date for payment enables a date to be established as the last date for giving notice of intention to withhold payment. Failure to pay by that date can also provide the contractor with a right to suspend performance.

The Act does not impose any credit period on the parties. They are free to agree any period at all. Once again though the object of the Act is to ensure that the parties do agree something so that there is no argument about when money should be paid.

If the parties fail to agree a method of establishing how much is to be paid, when a payment is due or a final date for payment, the Scheme will supplement their contract to achieve clarity.

10.4 Notice of amount to be paid

'S110–(2) Every construction contract shall provide for the giving of notice by a party not later than five days after the date on which a payment becomes due from him under the contract, or would have become due if –
  (a) the other party had carried out his obligations under the contract, and
  (b) no set-off or abatement was permitted by reference to any sum claimed to be due under one or more other contracts,
specifying the amount (if any) of the payment made or proposed to be made, and the basis on which that amount was calculated.

  (3) If or to the extent that a contract does not contain such

provision as is mentioned in subsection (1) or (2), the relevant provisions of the Scheme for Construction Contracts apply.'

In a further attempt to achieve clarity, the Act requires the payer to advise the payee of how much he is going to pay, within five days of a sum becoming due. The contract must provide for the giving of such a notice, and if it does not, the Scheme will fill the gap. The maximum period of five days has nothing to do with the final date for payment which may be some time off. It may be that there is in fact nothing to pay because the payee has not performed, or the payer may have a valid right of set-off or abatement in respect of matters arising under other contracts. Nevertheless the payer should still give notice.

It is not sufficient for the contract to require the payer to state the sum that he is going to pay. The basis of the calculation must be shown as well. If the value of the work is subject to some abatement because of defect, this should be made clear. The calculation that is required may involve a summary of the valuation of the works, a deduction by way of abatement, a statement of contra charges raised against both the current and previous payments, deduction of retention and discount and of course the sum of previous payments made. It should also deal with VAT. If there is a right to apply set-off in respect of other contracts (which is by no means always applicable), such set-off should also be detailed. The important question is: how much is going to be paid and how has it been calculated? On receipt of the notice, the payee should be entirely clear about the position. If the contract does not provide for this, the Scheme will do so.

The Act does not require the contract to contain any sanction to deal with a failure by the payee to give this notice. This has led to some confusion in dealing with contracts subject to the Scheme, to be discussed below.

10.5 Notice of intention to withhold payment

Section 111 requires a second notice to be given if the payer intends to withhold some or all of the payment that has fallen due:

'111-(1) A party to a construction contract may not withhold payment after the final date for payment of a sum due under the contract unless he has given an effective notice of intention to withhold payment.

The notice mentioned in section 110(2) may suffice as a notice of intention to withhold payment if it complies with the requirements of this section.

> (2) To be effective such a notice must specify –
>   (a) the amount proposed to be withheld and the ground for withholding payment, or
>   (b) if there is more than one ground, each ground and the amount attributable to it,
>
> and must be given not later than the prescribed period before the final date for payment.
>
> (3) The parties are free to agree what the prescribed period is to be.
> In the absence of such agreement, the period shall be that provided by the Scheme for Construction Contracts.
>
> (4) Where an effective notice of intention to withhold payment is given, but on the matter being referred to adjudication it is decided that the whole or part of the amount should be paid, the decision shall be construed as requiring payment not later than –
>   (a) seven days from the date of the decision, or
>   (b) the date which apart from the notice would have been the final date for payment,
>
> whichever is the later.'

Unlike the requirement for a notice to be given of the amount that is to be paid (under section 110) the requirement for a notice of intention to withhold does not depend on the construction contract. The parties are not required to include such a term in their contract. The Scheme for Construction Contracts is not invoked in order to fill in the gap if the requirement is not included in the contract. Whatever the contract says, a party to a construction contract may not withhold a payment that has become due unless an effective notice of intention to withhold has been given.

The notice given under section 110(2) may be effective as a notice of withholding under section 111, so long as the other requirements of section 111 are satisfied. This notice should have been given within five days of the money becoming due, by which time the ground(s) for withholding payment may have been clear. Notice of intention to withhold does not need to be given that early unless the parties have included a term to that effect in their contract. It is quite likely that the reason for wishing to withhold all or part of the payment will not have been apparent at the time of the earlier notice.

The requirements of the notice are simply that the amount to be withheld is stated and the ground given. If there is more than one ground, each should be stated with the amount relating to it. The contract may impose further requirements, although as it will probably have been drafted by the paying party it is unlikely to do so.

The notice must be given not later than the prescribed period before the final date for payment. The final date for payment is a date that is either established by the express words of the contract as required by section 110(1), or alternatively has been imposed on the contract in default by the Scheme.

There is no requirement for any minimum or maximum period of notice. The parties can, if they wish, decide to require notice of intention to withhold payment to be given 28 days before the final date for payment. Similarly the period if agreed in the contract can be as little as one day. The important point is that the parties should have addressed the issue and decided what the period should be. If they have not done so, the Scheme will impose a period of seven days.

10.6 Right to suspend

The right to suspend performance of the contract for non-payment is another provision that does not rely either on the parties to include a term in the contract, or on the Scheme to do it for them. Section 112 of the Act provides:

> '112–(1) Where a sum due under a construction contract is not paid in full by the final date for payment and no effective notice to withhold payment has been given, the person to whom the sum is due has the right (without prejudice to any other right or remedy) to suspend performance of his obligations under the contract to the party by whom payment ought to have been made ('the party in default').
>
> (2) The right may not be exercised without first giving to the party in default at least seven days' notice of intention to suspend performance, stating the ground or grounds on which it is intended to suspend performance.
>
> (3) The right to suspend performance ceases when the party in default makes payment in full of the amount due.

> (4) Any period during which performance is suspended in pursuance of the right conferred by this section shall be disregarded in computing for the purposes of any contractual time limit, by the party exercising the right or by a third party, to complete any work directly or indirectly affected by the exercise of the right.
>
> Where the contractual time limit is set by reference to a date rather than a period, the date shall be adjusted accordingly.'

Contractors and subcontractors who have not been paid have always been tempted to suspend work as a means of encouraging the cash to flow, but conventional legal advice prior to May 1998 was that such action was dangerous. Without an express term in the contract giving a right to suspend, it would be a breach of contract to do so. The breach of payment terms would not itself have justified a breach in return. Failure to pay would be unlikely to amount to a repudiation of contract entitling the contractor to rescind.

Even if there was an express or hard to find implied term giving a right to suspend for non-payment, it would be necessary to establish that a payment was due. Without a clear means of establishing whether a payment was in fact due, and with the possibility that unexpected contra-charges might be raised, this test could be difficult. The Act has now introduced into all construction contracts the right to suspend. Furthermore, the other provisions of the Act should operate to make it much clearer whether or not a payment is due. Contra-charges cannot be brought into the account by surprise, as there can be no withholding of a payment unless a proper notice has been given under section 111. If the term required by section 110 has been complied with, the sum to which the payee is entitled should also be clear well in advance of the final date for payment.

Section 112 has certainly made suspension a much less dangerous route to follow, but there are still serious difficulties. The right to suspend cannot be exercised without giving at least seven days' notice of intention to suspend. This may be a very expensive period, and the contractor will be obliged to continue his work. If he chooses to reduce resources during that period he may find that payment is made at the end of the seven days but he has dropped several days behind programme. There will be no extension of time, and he will have to accelerate or face the consequences of being late. On the other hand he may have serious doubts that he is going to be paid at all, and not want to increase his exposure.

Once suspension has started, the contractor has some relief from the programme consequences. Subsection 4 effectively gives the contractor an entitlement to an extension of time for the period of the suspension. But there is no entitlement to further time for remobilisation of resources in order to return to site and restart. Moreover there is no statutory right to payment of the cost that has been incurred in leaving site and returning.

The rather limited right of suspension may be enlarged by specific provision in the contract. The fact that there is a statutory right to suspend after giving seven days' notice does not prevent the parties from agreeing that there should be a contractual right to suspend with less notice. The parties might also agree that there should be an entitlement to an extension of time to cover remobilisation, and perhaps a payment of loss and expense (or similar) incurred in the suspension and remobilisation exercise. These two last points have been taken up by the JCT in the 1998 editions of the Standard Forms of Contract. They are however private contractual matters and not statutory rights.

It may also be possible for a contractor to argue that non-payment is a breach of contract, and that it must have been within the contemplation of the defaulting employer that the contractor would exercise his statutory right to suspend. The expense of suspension and remobilisation might then be claimed as damages arising out of the breach.

10.7 Conditional payment provisions

The 'pay when paid' clause was perhaps the one common feature of construction subcontracts that most offended the party who was hoping to be paid. Main contractors' terms of subcontract would be branded as aggressive if they contained such a clause and might be assumed benign if they did not. There was surprisingly little case law dealing with them, perhaps because main contractors were reluctant to allow their clauses to be tested in court in case they were declared ineffective. There were many different versions, some designed to have a pay when paid effect without being obvious, and some drafted with added sophistication in the hope that they would prove effective. The object of course was to pass the risk of non-payment by an impecunious employer down to the subcontractor, or at least to share that risk with the subcontractor. A secondary object was to protect the main contractor's cash flow. Even if the employer was able to pay, the main contractor did not wish to pay out the value of the work to the subcontractors until he had been paid.

The subcontractor felt that this was fundamentally unfair. He had no direct contractual relationship with the employer, except perhaps through a collateral warranty dealing with quality of workmanship or design. He had no means of ensuring that payment would be made by the employer or of pressing him for payment within the agreed credit period. Although it had become politically incorrect to advance the opposite argument, main contractors also had a point. The process of construction on a large site was to some degree a partnership between all the contractors involved. The main contractor was a conduit through whom the benefits of a project would be passed down to the subcontractors, but relatively little of the profit would remain in his hands. The most profitable trade on the site might be one of the specialists, such as mechanical and electrical engineering. There was some justice in asking that specialist to accept a portion of risk.

The draftsmen of the Act have taken a middle route in dealing with the pay when paid clause. Section 113 (1) provides:

'113–(1) A provision making payment under a construction contract conditional on the payer receiving payment from a third person is ineffective, unless that third person, or any other person payment by whom is under the contract (directly or indirectly) a condition of payment by that person, is insolvent.'

In simple cases the effect of this section is quite clear. A traditional 'pay when paid' or 'pay if paid' clause in a subcontract is ineffective. The main contractor will not be able to delay or avoid payment to his subcontractor on the strength of such a clause, simply because the employer has not paid. It is as if the clause was not there at all, and the normal payment provisions of the subcontract will apply.

If however the employer has not paid because he is insolvent, as defined by the Act, a traditional 'pay when paid' clause will be effective. The main contractor will be able to avoid paying the subcontractor. In this case the main contractor's position is stronger than before the Act, because there is no longer any doubt about the efficacy of the clause.

The position becomes rather less clear when there are other parties involved. The popular understanding of the section is that if there is an insolvency further up the chain of payment than described above, that insolvency will be sufficient to make the conditional payment provisions in all the subsequent contracts effective. If the employer fails to pay because he is insolvent, the pay when paid clause in the contract between the main contractor and the subcontractor will apply, and so will a pay when paid clause in a sub-subcontract.

This is not however quite what the section says. Looking at the situation described in the preceding paragraph from the point of view of the sub-subcontractor, he wishes to know whether the pay when paid clause in his contract with the first subcontractor works. The 'third person' who must pay the first subcontractor if the sub-subcontractor is to be paid is the main contractor. The main contractor is not insolvent. The insolvent party is the employer. Is the employer a 'person payment by whom is under the contract [i.e. the sub-subcontract] (directly or indirectly) a condition of payment by' the main contractor? It is unlikely that there will be a condition in the sub-subcontract dealing with payment by the main contractor to the first subcontractor. It may be argued that by inserting the words 'directly or indirectly' it is suggested that we are to consider the chain of payment in a simplistic way without worrying about which specific contract we are dealing with, but until the matter has been the subject of a court decision there must be considerable doubt.

A variant of the pay when paid clause in common use before the Act was the 'pay when certified' clause. In an attempt to secure the cash flow advantages of a pay when paid clause without the opprobrium that surrounded such clauses, main contractors would include a provision linking payment under the subcontract to certification under the main contract. If the main contract payment term was 14 days after the date of a certificate, the subcontract would provide for payment 21 days after the relevant main contract certificate. If the certificate was late, or was not given at all, the subcontractor would have to wait for payment.

There was some debate in the early days of the Act's passage through Parliament about whether section 113 should be expanded to make such a clause ineffective as well, but those who wished to do so were unsuccessful. It is argued by those who wish to champion the subcontractors' cause that pay when certified clauses are nevertheless ineffective, on the basis that the words 'receiving payment' in section 113 do not just mean 'receiving cash'. They argue that a certificate is chose in action or a debt, unless it is opened up, reviewed or revised in arbitration or litigation, as suggested in *Lubenham Fidelities and Investment Co v. South Pembrokeshire District Council* (1986) and *Costain Building and Civil Engineering Ltd v. Scottish Rugby Union plc* (1993). 'Pay when certified' therefore effectively becomes 'pay when paid' and is caught by the Act. Most commentators however consider that this is stretching the section too far.

It is also argued that whilst a 'pay when certified' clause may not be rendered ineffective by section 113, it does not provide an 'adequate mechanism for determining what payments become due under the contract and when', and that accordingly section 110 will intervene to impose the Scheme, rendering the 'pay when certified' clause redundant. The lack of an objective test for determining what is adequate makes it impossible to argue either case with certainty, but a provision that money will become due a specific number of days before or after an event such as the issue of a certificate under the main contract would seem capable of providing a mechanism that is workable and clear. To that extent it would seem to be adequate, if undesirable from the point of view of the subcontractor.

Subsections 113(2)–(5) define insolvency:

'113–(2) For the purposes of this section a company becomes insolvent –
- (a) on the making of an administration order against it under Part II of the Insolvency Act 1986,
- (b) on the appointment of an administrative receiver or a receiver or manager of its property under Chapter I of Part III of that Act, or the appointment of a receiver under Chapter II of that Part,
- (c) on the passing of a resolution for voluntary winding-up without a declaration of solvency under section 89 of that Act, or
- (d) on the making of a winding-up order under Part IV or V of that Act.

(3) For the purposes of this section a partnership becomes insolvent –
- (a) on the making of a winding-up order against it under any provision of the Insolvency Act 1986 as applied by an order under section 420 of that Act, or
- (b) when sequestration is awarded on the estate of the partnership under section 12 of the Bankruptcy (Scotland) Act 1985 or the partnership grants a trust deed for its creditors.

(4) For the purposes of this section an individual becomes insolvent –
- (a) on the making of a bankruptcy order against him under Part IX of the Insolvency Act 1986, or
- (b) on the sequestration of his estate under the Bankruptcy

(Scotland) Act 1985 or when he grants a trust deed for his creditors.

(5) A company, partnership or individual shall also be treated as insolvent on the occurrence of any event corresponding to those specified in subsection (2), (3) or (4) under the law of Northern Ireland or of a country outside the United Kingdom.'

It should be noted that this list is not the same as the list of circumstances included as 'insolvency' in many contracts. For example, the standard JCT forms of main contract include the making of a composition or arrangement with creditors. Nevertheless most forms of formal insolvency procedures are covered, and there is a wide catch-all provision to cover similar situations for parties resident outside the UK or subject to insolvency procedures in other countries.

Subsection (6) explains what happens if the contractual arrangement has been declared ineffective by the earlier subsections:

'113–(6) Where a provision is rendered ineffective by subsection (1), the parties are free to agree other terms for payment.
In the absence of such agreement, the relevant provisions of the Scheme for Construction Contracts apply.'

This suggests a rather unlikely procedure. The pay when paid clause having failed, the parties are free to agree other terms. By that time the subcontractor will be pressing for payment and the main contractor will be trying to avoid paying. In practice there will be no agreement, and consequently the Scheme will impose a payment mechanism to fill the gap. Of course there will be no need for any reference to the Scheme if the pay when paid clause was merely an overriding limitation on payment with a full compliant mechanism in place beneath it.

10.8 The Scheme

Whereas the Scheme provides a complete set of rules for adjudication which are imposed on any contract that does not fully comply with the Act's requirements for adjudication, the Scheme's provisions regarding payment are independent terms. If the contract's payment provisions partially comply with the Act's requirements those provisions will remain effective. The Scheme will replace only non-compliant or non-existent provisions.

The Scheme's provisions are set out in Part II of the statutory instrument.

10.8.1 Entitlement to and amount of stage payments

Paragraphs 1 and 2 of Part II of the Scheme provide the basis for establishing the entitlement to stage payments and the amount of them:

'1. Where the parties to a relevant construction contract fail to agree –
   (a) the amount of any instalments or stage or periodic payment for any work under the contract, or
   (b) the intervals at which, or circumstances in which, such payments become due under the contract, or
   (c) both of the matters mentioned in sub-paragraphs (a) and (b) above,

the relevant provisions of paragraphs 2 to 4 below shall apply.

2–(1) The amount of any payment by way of instalments or stage or periodic payments in respect of a relevant period shall be the difference between the amount determined in accordance with sub-paragraph (2) and the amount determined in accordance with sub-paragraph (3).

   (2) The aggregate of the following amounts –
      (a) an amount equal to the value of any work performed in accordance with the relevant construction contract during the period from the commencement of the contract to the end of the relevant period (including any amount calculated in accordance with sub-paragraph (b))
      (b) where the contract provides for payment for materials, an amount equal to the value of any materials manufactured on site or brought onto site for the purposes of the works during the period from the commencement of the contract to the end of the relevant period, and
      (c) any other amount or sum which the contract specifies shall be payable during or in respect of the period from the commencement of the contract to the end of the relevant period.

> (3) The aggregate of any sums which have been paid or are due for payment by way of instalments, stage or periodic payments during the period from the commencement of the contract to the end of the relevant period.
>
> (4) An amount calculated in accordance with this paragraph shall not exceed the difference between –
> (a) the contract price, and
> (b) the aggregate of the instalments or stage or periodic payments which have become due'

Definitions of terms used in these and the rest of Part II of the Scheme are set out in paragraph 12. 'Relevant construction contract' is defined as:

> 'any construction contract other than one –
> (a) which specifies that the duration of the work is estimated to be less than 45 days, or
> (b) in respect of which the parties agree that the duration of the work is estimated to be less than 45 days'

Section 109 of the Act provided that a party to a construction contract is entitled to payment by instalments etc. unless the above exclusions apply. It should be noted that there must be a specific mention of the duration being less than 45 days, or clear agreement to that effect, before the contract is taken out of the definition. In case of doubt therefore the right to stage payments will be included. Contracts that are within the exception remain construction contracts, subject to the Act and possibly the Scheme, but there is no statutory right to stage payments or similar.

Having thus established the right to stage payments, the Scheme then sets out a formula for calculating the amounts and when they are to be paid.

The first calculation required is the value of work. 'Work' is defined in paragraph 12 as 'any of the work or services mentioned in section 104 of the Act'. It includes only construction operations and services relating to such operations. Accordingly if the contract also covers other matters, the stage payments calculated and payable under the Scheme will not include such other matters.

'Value of work' is defined in paragraph 12 as:

> 'an amount determined in accordance with the construction contract under which the work is performed or where the contract contains no such provision, the cost of any work performed

in accordance with that contract together with an amount equal to any overhead or profit included in the contract price.'

If the contract contains any method of valuing the work, that method will be used to calculate the stage payment, but if the contract does not contain any such method, the payment is calculated on the basis of the cost of the work plus overhead and profit. Clearly therefore it is important to the paying party that a method should be established clearly in the contract. There is a protection in subparagraph (4) against the contract price being exceeded by the use of the cost plus method of valuation, but a stage payment valued on this basis may be substantially greater than the proper proportion of the contract price.

This first stage of calculation requires the evaluation of the work from the start of the contract to 'the end of the relevant period'. That term is defined in paragraph 12 as:

'a period which is specified in, or is calculated by reference to the construction contract or where no such period is so specified or is so calculable, a period of 28 days.'

If then the contract specifies that stage payments shall be payable every three months, that is the period that will be used for the calculation under the Scheme. There is no requirement for any particular period to be agreed, and any agreement will be respected by the Scheme. If however there is no agreement, stage payments will be due every 28 days. The calculation does not require consideration just of the work carried out since the last valuation. The total value to date is calculated.

Having calculated 'the value of work', the next stage is to calculate the value of materials, providing that the contract provides for payment for materials. Materials that have already been incorporated in the work will have been valued under the preceding paragraph. If there is no contractual entitlement to payment in respect of materials not yet incorporated, they will not be valued under this paragraph. There is no entitlement under this paragraph to payment for off-site materials. There is no definition of value of materials, and it is not clear whether the 'cost plus' basis that is the default method for valuing work applies. Once again, the valuation is to cover the whole period from commencement.

Finally there is a sweep-up paragraph, requiring the addition of any other sum payable under the contract. This would include any express entitlement to payment in respect of off-site materials, one-off design fees, etc.

Having calculated the total payable under the contract to date, the calculation then requires the subtraction of all sums that have been paid or have become due for payment from the start of the contract up to the end of the relevant period; the sum now being calculated has not of course yet become payable. The subtraction gives the net sum payable in respect of the relevant period.

There is then a safeguard against overpayment. The amount of a stage payment can never exceed the 'contract price', which is defined in paragraph 12 as 'the entire sum payable under the construction contract in respect of the work'. If therefore there has been a series of calculations based on cost plus overhead and profit which apparently values the work at rather more than the contract price there is a cap preventing the contractor from being paid the excess. 'Contract price' does not necessarily mean the sum that was agreed at the start – it includes the price of variations, and allowances for fluctuations and the like.

10.8.2 Dates for payment

The Act deals with two important dates: the date when a payment becomes due and the final date for payment. The difference was discussed above at section 10.3. The Scheme has to deal with both, and where a contract is to include stage payments (i.e. a contract expected to last more than 45 days) there has to be provision both for stage payments and the final payment.

Paragraphs 3–7 of the Scheme deal with the dates on which payments become due.

> '(3) Where the parties to a construction contract fail to provide an adequate mechanism for determining either what payments become due under the contract, or when they become due for payment, or both, the relevant provisions of paragraphs 4 to 7 shall apply.
>
> (4) Any payment of a kind mentioned in paragraph 2 above shall become due on whichever of the following dates occurs later –
>   (a) the expiry of 7 days following the relevant period mentioned in paragraph 2(1) above, or
>   (b) the making of a claim by the payee.
>
> (5) The final payment under a relevant construction contract, namely the payment of an amount equal to the difference (if any) between –

(a) the contract price, and
(b) the aggregate of any instalment or stage or periodic payments which have become due under the contract,

shall become due on the expiry of

(a) 30 days following completion of the work, or
(b) the making of a claim by the payee,

whichever is the later.

(6) Payment of the contract price under a construction contract (not being a relevant construction contract) shall become due on

(a) the expiry of 30 days following the completion of the work, or
(b) the making of a claim by the payee

whichever is the later.

(7) Any other payment under a construction contract shall become due

(a) on the expiry of 7 days following the completion of the work to which the payment relates, or
(b) the making of a claim by the payee

whichever is the later.'

Once again it must be remembered that these provisions are only relevant if the contract fails to provide an adequate mechanism. Under the Act the parties were free to agree any terms they liked to cover what payments were to become due and when they were to become due, subject only to the overriding requirement in a contract for 45 days or more for there to be stage payments of some sort. These paragraphs can only come into effect if the parties failed to do that.

Paragraph 4 deals with stage payments. It does not apply either to the final payment in a contract where there have been stage payments, nor does it apply to a contract which is not required to provide for stage payments because it was not anticipated that the work would last for 45 days. The stage payments will become due seven days after the expiry of each relevant period, discussed above, or on the making of a claim by the payee, whichever is the later.

It is therefore first necessary to consider when the relevant period expires. The contract may establish this. If it does not, then the first relevant period will expire 28 days after commencement of the contract. This does not necessarily mean commencement of work on site. If the work is the execution of groundworks, commencement probably will be the day that the contractor first goes to the site, but if the work includes design services the commencement will be the start of the performance of those services.

On the expiry of the first relevant period the second relevant period starts. This will be the case even if work is not continuous. There may well be gaps during which no work is carried out at all. Nevertheless relevant periods will continue to tick by.

The stage payment does not necessarily become due at the end of the relevant period, although the contract may provide that it does. If the contract is silent on the point, the stage payment will become due seven days later, or on the making of a claim by a payee. The later of these two dates is the date that payment becomes due. If therefore seven days pass without any claim being made by the payee, no payment is due until the claim is made. If the claim is presented within the seven day period, the stage payment becomes due at the end of that period. 'Claim by the payee' is defined in paragraph 12, and if the payee submits an application that does not satisfy the requirements of the definition the stage payment will not become due. The definition reads thus:

'a written notice given by the party carrying out work under a construction contract to the other party specifying the amount of any payment or payments which he considers to be due and the basis on which it is, or they are calculated.'

A simple statement that a figure is due will not be sufficient. The basis of calculation must be given. It is not however a requirement that the claim should be correct. A claim for a wildly exaggerated sum, with a clearly incorrect calculation, will be enough to ensure that a payment becomes due. The payment that becomes due will not be the sum for which application has been made, but the date that it becomes due will be established. It is clearly important to any prospective payee that a claim for payment is submitted as soon as possible after the end of the relevant period, or even perhaps a day or two before. This will ensure that the payment becomes due at the expiry of the seven days.

The final payment (not to be confused with the 'final date for payment', to be considered later) under a contract that has had stage payments is dealt with by paragraph 5 of Part II of the Scheme. The final payment is the total of the contract price, computed to take account of all omissions, additions and other sums due from one party to the other under the contract, less all the stage payments that have become due during the course of the work. It becomes due, subject of course to other provisions of the contract, 30 days after completion of the work or on the making of a claim by the payee, whichever is the later. There is no allowance made here for the sophistication of practical completion, defects periods, certification of making good and the like. If parties wish to provide for such things they must include express conditions in their contracts. Whether or not the work is complete, and when completion was achieved, will continue to be major sources of dispute.

Contracts that specify that the duration of the work will be less than 45 days or where the parties have agreed that the duration is estimated to be less than 45 days, are not subject to the requirement for stage payments, and unless the contract provides for interim payments there is no entitlement to payment until the end of the work. Paragraph 6 covers these contracts. The one payment is due 30 days after the completion of the work, or on the making of a claim by the payee, whichever is the later. The claim by the payee is subject to the same requirements as those discussed above.

Paragraph 7 sweeps up any payments that have not been caught by the previous paragraphs. Any such entitlement will be treated in the same way as a stage payment and will become due seven days after the completion of the work to which it relates or on the making of a claim, whichever is the later.

10.8.3 Final date for payment

The final date for payment is significant as the last date when the payee can expect to receive payment, the date on which the payee can give notice of intention to suspend performance and as the basis for establishing the latest date for the payer to give a notice of intention to withhold all or part of the payment. Paragraph 8 deals with a position where no final date for payment has been agreed:

'8-(1) Where the parties to a construction contract fail to provide a final date for payment in relation to any sum which becomes due under a construction contract, the provisions of this paragraph shall apply.

(2) The final date for the making of any payment of a kind mentioned in paragraphs 2, 5, 6 or 7 shall be 17 days from the date that payment becomes due.'

This is a simple statement. The final date for payment is 17 days after the payment becomes due and is not directly related to the valuation date or the end of the relevant period. If the payee has delayed putting in a claim for payment, so that the 'payment due' date is later than seven days after the end of the period, the final date for payment will also be extended.

10.8.4 Notice specifying amount of payment

Section 110(2) of the Act requires every construction contract to provide for the giving of a notice not later than five days after a payment becomes due of the amount of that payment and the basis of its calculation. If the contract fails to make such provision, paragraph 9 of the Scheme applies:

> '9. A party to a construction contract shall, not later than 5 days after the date upon which any payment –
> (a) becomes due from him, or
> (b) would have become due, if
>   (i) the other party had carried out his obligations under the contract, and
>   (ii) no set-off or abatement was permitted by reference to any sum claimed to be due under one or more other contracts,
> give notice to the other party to the contract specifying the amount (if any) of the payment he has made or proposed to make, specifying to what the payment relates and the basis on which that amount is calculated.'

This process is something like a traditional certification, but the 'certificate' is not to be given by a quasi-independent third party, such as the engineer or the architect, but by the paying party himself. It is to be given not more than five days after the payment became due, which as we have seen may not be until the payee has made a claim setting out in some detail what he believes he is entitled to be paid.

If no payment has become due, perhaps because the payee has not done any work, or because there are cross-claims that the payer is contractually entitled to set off against the value of work carried out, the payer must still give his notice explaining why nothing is going to be paid.

It will not be sufficient merely to state the sum that is going to be paid. Details must be provided showing what the payment is for and the basis of the calculation. This may be substantially more

Payment 227

information than was common before the Act came into force. In a typical subcontract the following will probably need to be set out clearly in order to comply with this paragraph:

- Gross valuation of work to date, with breakdown
- Retention, if applicable, at subcontract rate
- Main contractor's discount, if applicable
- Sums payable as previous stage payments
- Sums withheld from previous stage payments following notice(s) of intention to withhold, unless such sums are now being released
- VAT as applicable
- Resulting sum to be paid.

The intention is of course that the payer should give sufficient information to enable the payee to understand exactly what he is being paid and why, so that if he considers that he is not being paid his proper entitlement he can raise the issue at a comparatively early stage. Hopefully that matter can be resolved by discussion, but if necessary adjudication is available without any serious delay. It is curious that this intention is not supported by any sanction for failure to give a notice in accordance with this paragraph. There is nothing in the Scheme that provides for anything to happen if the payer gives no notice at all, let alone a notice that is not fully detailed.

This was not the original intention of the draftsmen of the Scheme. In early drafts it was suggested that the payee would present his application for payment, and unless the payer countered the application with a detailed statement of what he thought was the appropriate amount and why, the payee's application would be conclusive of the amount that would have to be paid, subject of course to the right of set-off on giving notice as required by section 111. This proposal was well publicised, for example in the consultation paper *Making the Scheme for Construction Contracts*, and it was widely expected that this would be the position in all contracts to which the Scheme applied.

This original intention was abandoned, but it is commonly argued in adjudication proceedings that if the payer has failed to give a notice specifying the payment that he intends to make, the claim made by the payee should be taken as being conclusive as to the sum that should be paid as a stage payment. Specific terms of the contract may give this argument a proper basis (as for example in the JCT With Contractor's Design contract) but there is nothing in the Scheme to support it. Nevertheless, uncertainty on this point is widespread, particularly in view of the very clear sanction for failure to give a notice of intention to withhold under section 111 of the Act and paragraph 10 of Part II of the Scheme. Discussion of the point continues in the next section.

10.8.5 Notice of intention to withhold payment

Section 111 of the Act has direct effect, and does not need a contractual provision to require a party to give notice of intention to withhold a payment. The requirement is imposed by the statute as are the required contents of the notice and the consequences of failure to give the notice. The only matter that is left for agreement between the parties is the length of notice that is required. If that point is not covered in the contract the Scheme provides:

> '10. Any notice of intention to withhold payment mentioned in Section 111 of the Act shall be given not later than the prescribed period, which is to say not later than 7 days before the final date for payment determined either in accordance with the construction contract, or where no such provision is made in the contract, in accordance with paragraph 8 above.'

The notice must be given not less than seven days before the final date for payment, which itself is established either by the contract or in default by paragraph 8 of Part II of the Scheme.

If the payer has not given notice of intention to withhold in due time he will not be able to withhold the payment or any part of it. This clear sanction for failure to give a notice required by the Act is in stark contrast to the absence of any sanction for failure to give the notice of the amount of the proposed payment within five days from the date the payment became due. The apparent requirement for certainty has led to a search for some greater significance to attach to the earlier notice than can be justified by the words of the Act and the Scheme.

The judgment of Judge Bowsher QC in *Northern Developments (Cumbria) Ltd v. J. & J. Nichol* (January 2000) has led to some confusion on this point. He reviewed the statutory provisions regarding payment and came to this conclusion:

> 'The Act by section 111 imposes on the parties a direct requirement that the paying party may not withhold a payment after the due date for payment unless he has given an effective notice of

intention to withhold payment, That seems to me to have a direct bearing on the ambit of any dispute to be heard by an adjudicator. Section 110 requires that the contract must require that within 5 days of any sum falling due under the contract, the paying party must give a statement of the amount due or of what would be due if the payee had performed the contract. Section 111 provides that no deduction can be made after the final date for payment unless the paying party has given notice of intention to withhold payment. The intention of the statute is clearly that if there is to be a dispute about the amount of the payment required by section 111, that dispute is to be mentioned in a notice of intention to withhold payment not later than 5 days after the due date for payment... For the temporary striking of balances which are contemplated by the Act, there is to be no dispute about any matter not raised in the notice of intention to withhold payment.'

With respect to the learned judge, there are several points within the passage set out above which are misleading. Firstly, section 110 does not require 'that the contract must require that within five days of any sum falling due under the contract, the paying party must give a statement of the amount due or of what would be due if the payee had performed the contract'. The section deals with the timing of the notice. The notice is to be given no later than five days after the date on which the payment becomes due, *or would have become due if the other party had carried out its obligations* etc. There is no need to state what amount would have become due if the payee had performed.

There is then a suggestion that if there is any dispute about the amount to be paid, the dispute is to be mentioned in a notice of intention to withhold given not later than five days after the due date for payment. There are many potential disputes that will not be mentioned in a notice of intention to withhold. A dispute about the value of work done will not be covered by any notice of intention to withhold. If the first of the two statements has been given, there may well be a clear dispute on the valuation, but as has been stressed above there is no sanction for failure to give that notice.

Furthermore, a notice of intention to withhold need not be given as early as five days after the due date for payment. If the Scheme is operating in all respects, the contract having completely failed to provide any of the Act's requirements, the final date for payment is 17 days after the payment due date, and the latest date for notice of intention to withhold is seven days before that, or 10 days after the payment due date.

In fact in the *Northern Developments* case, the adjudicator had correctly interpreted the position on this point. In a letter to the parties he had said:

> 'As for the lack of notices pursuant to section 110(2) of the Act, I considered that as the Scheme is silent as to the consequences of failure to comply and furthermore as the value of the works had to be that properly carried out, then as stated previously, I decided that the question of defects and their value could be dealt with by me.'

If an adjudicator, or indeed a judge or arbitrator, is asked to decide how much is properly payable he must go back to first principles and enquire into the value of the works, established in accordance with the contract.

Judge Bowsher revisited the question of valuation and the relationship between sections 110 and 111 of the Act in *Whiteways Contractors (Sussex) Ltd* v. *Impresa Castelli Construction UK Ltd* (August 2000):

> 'Of course, in considering a dispute, an adjudicator will make his own valuation of the claim before him and in doing so, he may abate the claim in respects not mentioned in the notice of intention to withhold payment. But he ought not to look into abatements outside the four corners of the claim unless they have been mentioned in a notice of intention to withhold payment. So, to take a hypothetical example, if there is dispute about valuation 10, the adjudicator may make his own valuation of the matters referred to in valuation 10 whether or not they are referred to him specifically in a notice of intention to withhold payment. But it would be wrong for him to enquire into an alleged overvaluation on valuation 6, whether the paying party alleges abatement or set-off, unless the notice of intention to withhold payment identified that as a matter of dispute.'

If the dispute is about valuation 10, then a claim that valuation 6 was overvalued and a credit is due back is in reality a claim for a set-off, not a claim for an abatement.

10.8.6 Prohibition of conditional payment provisions

Section 113 of the Act provides that a conditional payment provision (a 'pay when paid' or 'pay if paid' clause) is ineffective except in certain circumstances involving insolvency). This was discussed in section 10.7 of this chapter. The Scheme also contains a provision relating to conditional payments, providing what is to happen if section 113 of the Act has operated and effectively destroyed the payment mechanism in the contract that relied on the conditional payment clause.

Paragraph 11 of Part II of the Scheme provides:

'11. Where a provision making payment under a construction contract conditional on the payer receiving payment from a third person is ineffective as mentioned in section 113 of the Act, and then parties have not agreed other terms for payment, the relevant provisions of –
(a) paragraphs 2, 4, 5, 7, 8, 9 and 10 shall apply in the case of a relevant construction contract, and
(b) paragraphs 6, 7, 8, 9 and 10 shall apply in the case of any other construction contract.'

A 'relevant construction contract' is a contract that specifies that the work is to be of less than 45 days duration or in respect of which the parties agree that the work is estimated to be of less than 45 days duration. The effect of this paragraph is simply to ensure that the Scheme will apply not only where no agreement has been reached providing an adequate mechanism for dealing with payments, but also where the mechanism has been declared ineffective.

APPENDIX 1
HOUSING GRANTS, CONSTRUCTION AND REGENERATION ACT 1996

Part II
Construction contracts
Introductory provisions

Construction contracts.

104.–(1) In this Part a "construction contract" means an agreement with a person for any of the following–
  (a) the carrying out of construction operations;
  (b) arranging for the carrying out of construction operations by others, whether under sub-contract to him or otherwise;
  (c) providing his own labour, or the labour of others, for the carrying out of construction operations.

(2) References in this Part to a construction contract include an agreement–
  (a) to do architectural, design, or surveying work, or
  (b) to provide advice on building, engineering, interior or exterior decoration or on the laying-out of landscape,
in relation to construction operations.

(3) References in this Part to a construction contract do not include a contract of employment (within the meaning of the Employment Rights Act 1996).

(4) The Secretary of State may by order add to, amend or repeal any of the provisions of subsection (1), (2) or (3) as to the agreements which are construction contracts for the purposes of this Part or are to be taken or not to be taken as included in references to such contracts.
No such order shall be made unless a draft of it has been laid before and approved by a resolution of each House of Parliament.

(5) Where an agreement relates to construction operations and other matters, this Part applies to it only so far as it relates to construction operations.
An agreement relates to construction operations so far as it makes provision of any kind within subsection (1) or (2).

(6) This Part applies only to construction contracts which–
  (a) are entered into after the commencement of this Part, and
  (b) relate to the carrying out of construction operations in England, Wales or Scotland.

(7) This Part applies whether or not the law of England and Wales or Scotland is otherwise the applicable law in relation to the contract.

105.–(1) In this Part "construction operations" means, subject as follows, operations of any of the following descriptions– *Meaning of "construction operations".*
  (a) construction, alteration, repair, maintenance, extension, demolition or dismantling of buildings, or structures forming, or to form, part of the land (whether permanent or not);
  (b) construction, alteration, repair, maintenance, extension, demolition or dismantling of any works forming, or to form, part of the land, including (without prejudice to the foregoing) walls, roadworks, power-lines, telecommunication apparatus, aircraft runways, docks and harbours, railways, inland waterways, pipe-lines, reservoirs, water-mains, wells, sewers, industrial plant and installations for purposes of land drainage, coast protection or defence;
  (c) installation in any building or structure of fittings forming part of the land, including (without prejudice to the foregoing) systems of heating, lighting, air-conditioning, ventilation, power supply, drainage, sanitation, water supply or fire protection, or security or communications systems;
  (d) external or internal cleaning of buildings and structures, so far as carried out in the course of their construction, alteration, repair, extension or restoration;
  (e) operations which form an integral part of, or are preparatory to, or are for rendering complete, such operations as are previously described in this subsection, including site clearance, earthmoving, excavation, tunnelling and boring, laying of foundations, erection, maintenance or dismantling of scaffolding, site restoration, landscaping and the provision of roadways and other access works;
  (f) painting or decorating the internal or external surfaces of any building or structure.

(2) The following operations are not construction operations within the meaning of this Part–
  (a) drilling for, or extraction of, oil or natural gas;
  (b) extraction (whether by underground or surface working) of minerals; tunnelling or boring, or construction of underground works, for this purpose;
  (c) assembly, installation or demolition of plant or machinery, or erection or demolition of steelwork for the purposes of sup-

porting or providing access to plant or machinery, on a site where the primary activity is–
>(i) nuclear processing, power generation, or water or effluent treatment, or
>(ii) the production, transmission, processing or bulk storage (other than warehousing) of chemicals, pharmaceuticals, oil, gas, steel or food and drink;

(d) manufacture or delivery to site of–
>(i) building or engineering components or equipment,
>(ii) materials, plant or machinery, or
>(iii) components for systems of heating, lighting, air-conditioning, ventilation, power supply, drainage, sanitation, water supply or fire protection, or for security or communications systems,

except under a contract which also provides for their installation;

(e) the making, installation and repair of artistic works, being sculptures, murals and other works which are wholly artistic in nature.

(3) The Secretary of State may by order add to, amend or repeal any of the provisions of subsection (1) or (2) as to the operations and work to be treated as construction operations for the purposes of this Part.

(4) No such order shall be made unless a draft of it has been laid before and approved by a resolution of each House of Parliament.

Provisions not applicable to contract with residential occupier.

106.–(1) This Part does not apply–
(a) to a construction contract with a residential occupier (see below), or
(b) to any other description of construction contract excluded from the operation of this Part by order of the Secretary of State.

(2) A construction contract with a residential occupier means a construction contract which principally relates to operations on a dwelling which one of the parties to the contract occupies, or intends to occupy, as his residence.

In this subsection "dwelling" means a dwelling-house or a flat; and for this purpose–
>"dwelling-house" does not include a building containing a flat; and
>"flat" means separate and self-contained premises constructed or adapted for use for residential purposes and forming part of a building from some other part of which the premises are divided horizontally.

(3) The Secretary of State may by order amend subsection (2).

(4) No order under this section shall be made unless a draft of it has been laid before and approved by a resolution of each House of Parliament.

107.–(1) The provisions of this Part apply only where the construction contract is in writing, and any other agreement between the parties as to any matter is effective for the purposes of this Part only if in writing.

The expressions "agreement", "agree" and "agreed" shall be construed accordingly.

Provisions applicable only to agreements in writing.

(2) There is an agreement in writing–
  (a) if the agreement is made in writing (whether or not it is signed by the parties),
  (b) if the agreement is made by exchange of communications in writing, or
  (c) if the agreement is evidenced in writing.

(3) Where parties agree otherwise than in writing by reference to terms which are in writing, they make an agreement in writing.

(4) An agreement is evidenced in writing if an agreement made otherwise than in writing is recorded by one of the parties, or by a third party, with the authority of the parties to the agreement.

(5) An exchange of written submissions in adjudication proceedings, or in arbitral or legal proceedings in which the existence of an agreement otherwise than in writing is alleged by one party against another party and not denied by the other party in his response constitutes as between those parties an agreement in writing to the effect alleged.

(6) References in this Part to anything being written or in writing include its being recorded by any means.

Adjudication

108.–(1) A party to a construction contract has the right to refer a dispute arising under the contract for adjudication under a procedure complying with this section.

For this purpose "dispute" includes any difference.

Right to refer disputes to adjudication.

(2) The contract shall–
  (a) enable a party to give notice at any time of his intention to refer a dispute to adjudication;
  (b) provide a timetable with the object of securing the appointment of the adjudicator and referral of the dispute to him within 7 days of such notice;
  (c) require the adjudicator to reach a decision within 28 days of referral or such longer period as is agreed by the parties after the dispute has been referred;
  (d) allow the adjudicator to extend the period of 28 days by up to 14 days, with the consent of the party by whom the dispute was referred;
  (e) impose a duty on the adjudicator to act impartially; and

(f) enable the adjudicator to take the initiative in ascertaining the facts and the law.

(3) The contract shall provide that the decision of the adjudicator is binding until the dispute is finally determined by legal proceedings, by arbitration (if the contract provides for arbitration or the parties otherwise agree to arbitration) or by agreement.

The parties may agree to accept the decision of the adjudicator as finally determining the dispute.

(4) The contract shall also provide that the adjudicator is not liable for anything done or omitted in the discharge or purported discharge of his functions as adjudicator unless the act or omission is in bad faith, and that any employee or agent of the adjudicator is similarly protected from liability.

(5) If the contract does not comply with the requirements of subsections (1) to (4), the adjudication provisions of the Scheme for Construction Contracts apply.

(6) For England and Wales, the Scheme may apply the provisions of the Arbitration Act 1996 with such adaptations and modifications as appear to the Minister making the scheme to be appropriate.

For Scotland, the Scheme may include provision conferring powers on courts in relation to adjudication and provision relating to the enforcement of the adjudicator's decision.

Payment

Entitlement to stage payments.

109.-(1) A party to a construction contract is entitled to payment by instalments, stage payments or other periodic payments for any work under the contract unless–
  (a) it is specified in the contract that the duration of the work is to be less than 45 days, or
  (b) it is agreed between the parties that the duration of the work is estimated to be less than 45 days.

(2) The parties are free to agree the amounts of the payments and the intervals at which, or circumstances in which, they become due.

(3) In the absence of such agreement, the relevant provisions of the Scheme for Construction Contracts apply.

(4) References in the following sections to a payment under the contract include a payment by virtue of this section.

Dates for payment.

110.-(1) Every construction contract shall–
  (a) provide an adequate mechanism for determining what payments become due under the contract, and when, and
  (b) provide for a final date for payment in relation to any sum which becomes due.

The parties are free to agree how long the period is to be between the date on which a sum becomes due and the final date for payment.

(2) Every construction contract shall provide for the giving of notice by a party not later than five days after the date on which a payment becomes due from him under the contract, or would have become due if–
  (a) the other party had carried out his obligations under the contract, and
  (b) no set-off or abatement was permitted by reference to any sum claimed to be due under one or more other contracts,
specifying the amount (if any) of the payment made or proposed to be made, and the basis on which that amount was calculated.

(3) If or to the extent that a contract does not contain such provision as is mentioned in subsection (1) or (2), the relevant provisions of the Scheme for Construction Contracts apply.

111.–(1) A party to a construction contract may not withhold payment after the final date for payment of a sum due under the contract unless he has given an effective notice of intention to withhold payment. **Notice of intention to withhold payment.**

The notice mentioned in section 110(2) may suffice as a notice of intention to withhold payment if it complies with the requirements of this section.

(2) To be effective such a notice must specify–
  (a) the amount proposed to be withheld and the ground for withholding payment, or
  (b) if there is more than one ground, each ground and the amount attributable to it,
and must be given not later than the prescribed period before the final date for payment.

(3) The parties are free to agree what that prescribed period is to be.
In the absence of such agreement, the period shall be that provided by the Scheme for Construction Contracts.

(4) Where an effective notice of intention to withhold payment is given, but on the matter being referred to adjudication it is decided that the whole or part of the amount should be paid, the decision shall be construed as requiring payment not later than–
  (a) seven days from the date of the decision, or
  (b) the date which apart from the notice would have been the final date for payment,
whichever is the later.

112.–(1) Where a sum due under a construction contract is not paid in full by the final date for payment and no effective notice to withhold payment has been given, the person to whom the sum is due has the right (without prejudice to any other right or remedy) **Right to suspend performance for non-payment.**

to suspend performance of his obligations under the contract to the party by whom payment ought to have been made ("the party in default").

(2) The right may not be exercised without first giving to the party in default at least seven days' notice of intention to suspend performance, stating the ground or grounds on which it is intended to suspend performance.

(3) The right to suspend performance ceases when the party in default makes payment in full of the amount due.

(4) Any period during which performance is suspended in pursuance of the right conferred by this section shall be disregarded in computing for the purposes of any contractual time limit the time taken, by the party exercising the right or by a third party, to complete any work directly or indirectly affected by the exercise of the right.

Where the contractual time limit is set by reference to a date rather than a period, the date shall be adjusted accordingly.

Prohibition of conditional payment provisions.

113.–(1) A provision making payment under a construction contract conditional on the payer receiving payment from a third person is ineffective, unless that third person, or any other person payment by whom is under the contract (directly or indirectly) a condition of payment by that third person, is insolvent.

(2) For the purposes of this section a company becomes insolvent–
  (a) on the making of an administration order against it under Part II of the Insolvency Act 1986,
  (b) on the appointment of an administrative receiver or a receiver or manager of its property under Chapter I of Part III of that Act, or the appointment of a receiver under Chapter II of that Part,
  (c) on the passing of a resolution for voluntary winding-up without a declaration of solvency under section 89 of that Act, or
  (d) on the making of a winding-up order under Part IV or V of that Act.

(3) For the purposes of the section a partnership becomes insolvent–
  (a) on the making of a winding-up order against it under any provision of the Insolvency Act 1986 as applied by an order under section 420 of that Act, or
  (b) when sequestration is awarded on the estate of the partnership under section 12 of the Bankruptcy (Scotland) Act 1985 or the partnership grants a trust deed for its creditors.

(4) For the purposes of this section an individual becomes insolvent–
  (a) on the making of a bankruptcy order against him under Part IX of the Insolvency Act 1986, or

(b) on the sequestration of his estate under the Bankruptcy (Scotland) Act 1985 or when he grants a trust deed for his creditors.

(5) A company, partnership or individual shall also be treated as insolvent on the occurrence of any event corresponding to those specified in subsection (2), (3) or (4) under the law of Northern Ireland or of a country outside the United Kingdom.

(6) Where a provision is rendered ineffective by subsection (1), the parties are free to agree other terms for payment.

In the absence of such agreement, the relevant provisions of the Scheme for Construction Contracts apply.

Supplementary provisions

114.–(1) The Minister shall by regulations make a scheme ("the Scheme for Construction Contracts") containing provision about the matters referred to in the preceding provisions of this Part. **The Scheme for Construction Contracts.**

(2) Before making any regulations under this section the Minister shall consult such persons as he thinks fit.

(3) In this section "the Minister" means–
  (a) for England and Wales, the Secretary of State, and
  (b) for Scotland, the Lord Advocate.

(4) Where any provisions of the Scheme for Construction Contracts apply by virtue of this Part in default of contractual provision agreed by the parties, they have effect as implied terms of the contract concerned.

(5) Regulations under this section shall not be made unless a draft of them has been approved by resolution of each House of Parliament.

115.–(1) The parties are free to agree on the manner of service of any notice or other document required or authorised to be served in pursuance of the construction contract or for any of the purposes of this Part. **Service of notices, &c.**

(2) If or to the extent that there is no such agreement the following provisions apply.

(3) A notice or other document may be served on a person by any effective means.

(4) If a notice or other document is addressed, pre-paid and delivered by post–
  (a) to the addressee's last known principal residence or, if he is or has been carrying on a trade, profession or business, his last known principal business address, or
  (b) where the addressee is a body corporate, to the body's registered or principal office,
it shall be treated as effectively served.

(5) This section does not apply to the service of documents for the purposes of legal proceedings, for which provision is made by rules of court.

(6) References in this Part to a notice or other document include any form of communication in writing and references to service shall be construed accordingly.

Reckoning periods of time.

116.–(1) For the purposes of this Part periods of time shall be reckoned as follows.

(2) Where an act is required to be done within a specified period after or from a specified date, the period begins immediately after that date.

(3) Where the period would include Christmas Day, Good Friday or a day which under the Banking and Financial Dealings Act 1971 is a bank holiday in England and Wales or, as the case may be, in Scotland, that day shall be excluded.

Crown application.

117.–(1) This Part applies to a construction contract entered into by or on behalf of the Crown otherwise than by or on behalf of Her Majesty in her private capacity.

(2) This Part applies to a construction contract entered into on behalf of the Duchy of Cornwall notwithstanding any Crown interest.

(3) Where a construction contract is entered into by or on behalf of Her Majesty in right of the Duchy of Lancaster, Her Majesty shall be represented, for the purposes of any adjudication or other proceedings arising out of the contract by virtue of this Part, by the Chancellor of the Duchy or such person as he may appoint.

(4) Where a construction contract is entered into on behalf of the Duchy of Cornwall, the Duke of Cornwall or the possessor for the time being of the Duchy shall be represented, for the purposes of any adjudication or other proceedings arising out of the contract by virtue of this Part, by such person as he may appoint.

APPENDIX 2
THE SCHEME FOR CONSTRUCTION CONTRACTS (ENGLAND AND WALES) REGULATIONS 1998

Statutory Instrument 1998 No. 649

The Secretary of State, in exercise of the powers conferred on him by sections 108(6), 114 and 146(1) and (2) of the Housing Grants, Construction and Regeneration Act 1996, and of all other powers enabling him in that behalf, having consulted such persons as he thinks fit, and draft Regulations having been approved by both Houses of Parliament, hereby makes the following Regulations:

Citation, commencement, extent and interpretation

1.-(1) These Regulations may be cited as the Scheme for Construction Contracts (England and Wales) Regulations 1998 and shall come into force at the end of the period of 8 weeks beginning with the day on which they are made (the "commencement date").

(2) These Regulations shall extend only to England and Wales.

(3) In these Regulations, "the Act" means the Housing Grants, Construction and Regeneration Act 1996.

The Scheme for Construction Contracts

2. Where a construction contract does not comply with the requirements of section 108(1) to (4) of the Act, the adjudication provisions in Part I of the Schedule to these Regulations shall apply.

3. Where–
  (a) the parties to a construction contract are unable to reach agreement for the purposes mentioned respectively in sections 109, 111 and 113 of the Act, or
  (b) a construction contract does not make provision as required by section 110 of the Act,

the relevant provisions in Part II of the Schedule to these Regulations shall apply.

4. The provisions in the Schedule to these Regulations shall be the Scheme for Construction Contracts for the purposes of section 114 of the Act.

SCHEDULE   Regulations 2, 3 and 4

THE SCHEME FOR CONSTRUCTION CONTRACTS

PART I–ADJUDICATION

Notice of Intention to seek Adjudication

1.-(1) Any party to a construction contract (the "referring party") may give written notice (the "notice of adjudication") of his intention to refer any dispute arising under the contract, to adjudication.

(2) The notice of adjudication shall be given to every other party to the contract.

(3) The notice of adjudication shall set out briefly–
- (a) the nature and a brief description of the dispute and of the parties involved,
- (b) details of where and when the dispute has arisen,
- (c) the nature of the redress which is sought, and
- (d) the names and addresses of the parties to the contract (including, where appropriate, the addresses which the parties have specified for the giving of notices).

2.–(1) Following the giving of a notice of adjudication and subject to any agreement between the parties to the dispute as to who shall act as adjudicator–
- (a) the referring party shall request the person (if any) specified in the contract to act as adjudicator, or
- (b) if no person is named in the contract or the person named has already indicated that he is unwilling or unable to act, and the contract provides for a specified nominating body to select a person, the referring party shall request the nominating body named in the contract to select a person to act as adjudicator, or
- (c) where neither paragraph (a) nor (b) above applies, or where the person referred to in (a) has already indicated that he is unwilling or unable to act and (b) does not apply, the referring party shall request an adjudicator nominating body to select a person to act as adjudicator.

(2) A person requested to act as adjudicator in accordance with the provisions of paragraph (1) shall indicate whether or not he is willing to act within two days of receiving the request.

(3) In this paragraph, and in paragraphs 5 and 6 below, an "adjudicator nominating body" shall mean a body (not being a natural person and not being a party to the dispute) which holds itself out publicly as a body which will select an adjudicator when requested to do so by a referring party.

3. The request referred to in paragraphs 2, 5 and 6 shall be accompanied by a copy of the notice of adjudication.

4. Any person requested or selected to act as adjudicator in accordance with paragraphs 2, 5 or 6 shall be a natural person acting in his personal capacity. A person requested or selected to act as an adjudicator shall not be an employee of any of the parties to the dispute and shall declare any interest, financial or otherwise, in any matter relating to the dispute.

5.-(1) The nominating body referred to in paragraphs 2(1)(b) and 6(1)(b) or the adjudicator nominating body referred to in paragraphs 2(1)(c), 5(2)(b) and 6(1)(c) must communicate the selection of an adjudicator to the referring party within five days of receiving a request to do so.

(2) Where the nominating body or the adjudicator nominating body fails to comply with paragraph (1), the referring party may–
   (a) agree with the other party to the dispute to request a specified person to act as adjudicator, or
   (b) request any other adjudicator nominating body to select a person to act as adjudicator.

(3) The person requested to act as adjudicator in accordance with the provisions of paragraphs (1) or (2) shall indicate whether or not he is willing to act within two days of receiving the request.

6.-(1) Where an adjudicator who is named in the contract indicates to the parties that he is unable or unwilling to act, or where he fails to respond in accordance with paragraph 2(2), the referring party may–
   (a) request another person (if any) specified in the contract to act as adjudicator, or
   (b) request the nominating body (if any) referred to in the contract to select a person to act as adjudicator, or
   (c) request any other adjudicator nominating body to select a person to act as adjudicator.

(2) The person requested to act in accordance with the provisions of paragraph (1) shall indicate whether or not he is willing to act within two days of receiving the request.

7.-(1) Where an adjudicator has been selected in accordance with paragraphs 2, 5 or 6, the referring party shall, not later than seven days from the date of the notice of adjudication, refer the dispute in writing (the "referral notice") to the adjudicator.

(2) A referral notice shall be accompanied by copies of, or relevant extracts from, the construction contract and such other documents as the referring party intends to rely upon.

(3) The referring party shall, at the same time as he sends to the adjudicator the documents referred to in paragraphs (1) and (2), send copies of those documents to every other party to the dispute.

8.-(1) The adjudicator may, with the consent of all the parties to those disputes, adjudicate at the same time on more than one dispute under the same contract.

(2) The adjudicator may, with the consent of all the parties to those disputes, adjudicate at the same time on related disputes under different contracts, whether or not one or more of those parties is a party to those disputes.

(3) All the parties in paragraphs (1) and (2) respectively may agree to extend the period within which the adjudicator may reach a decision in relation to all or any of these disputes.

(4) Where an adjudicator ceases to act because a dispute is to be adjudicated on by another person in terms of this paragraph, that adjudicator's fees and expenses shall be determined in accordance with paragraph 25.

9.-(1) An adjudicator may resign at any time on giving notice in writing to the parties to the dispute.

(2) An adjudicator must resign where the dispute is the same or substantially the same as one which has previously been referred to adjudication, and a decision has been taken in that adjudication.

(3) Where an adjudicator ceases to act under paragraph 9(1)-
  (a) the referring party may serve a fresh notice under paragraph 1 and shall request an adjudicator to act in accordance with paragraphs 2 to 7; and
  (b) if requested by the new adjudicator and insofar as it is reasonably practicable, the parties shall supply him with copies of all documents which they had made available to the previous adjudicator.

(4) Where an adjudicator resigns in the circumstances referred to in paragraph (2), or where a dispute varies significantly from the dispute referred to him in the referral notice and for that reason he is not competent to decide it, the adjudicator shall be entitled to the payment of such reasonable amount as he may determine by way of fees and expenses reasonably incurred by him. The parties shall be jointly and

severally liable for any sum which remains outstanding following the making of any determination on how the payment shall be apportioned.

10. Where any party to the dispute objects to the appointment of a particular person as adjudicator, that objection shall not invalidate the adjudicator's appointment nor any decision he may reach in accordance with paragraph 20.

11.–(1) The parties to a dispute may at any time agree to revoke the appointment of the adjudicator. The adjudicator shall be entitled to the payment of such reasonable amount as he may determine by way of fees and expenses incurred by him. The parties shall be jointly and severally liable for any sum which remains outstanding following the making of any determination on how the payment shall be apportioned.

(2) Where the revocation of the appointment of the adjudicator is due to the default or misconduct of the adjudicator, the parties shall not be liable to pay the adjudicator's fees and expenses.

Powers of the adjudicator

12. The adjudicator shall–
   (a) act impartially in carrying out his duties and shall do so in accordance with any relevant terms of the contract and shall reach his decision in accordance with the applicable law in relation to the contract; and
   (b) avoid incurring unnecessary expense.

13. The adjudicator may take the initiative in ascertaining the facts and the law necessary to determine the dispute, and shall decide on the procedure to be followed in the adjudication. In particular he may–
   (a) request any party to the contract to supply him with documents as he may reasonably require including, if he so directs, any written statement from any party to the contract supporting or supplementing the referral notice and any other documents given under paragraph 7(2),
   (b) decide the language or languages to be used in the adjudication and whether a translation of any document is to be provided and if so by whom,
   (c) meet and question any of the parties to the contract and their representatives,
   (d) subject to obtaining any necessary consent from a third party or parties, make such site visits and inspections as he considers appropriate, whether accompanied by the parties or not,
   (e) subject to obtaining any necessary consent from a third party or parties, carry out any tests or experiments,
   (f) obtain and consider such representations and submissions as

he requires, and, provided he has notified the parties of his intention, appoint experts, assessors or legal advisers,

(g) give directions as to the timetable for the adjudication, any deadlines, or limits as to the length of written documents or oral representations to be complied with, and

(h) issue other directions relating to the conduct of the adjudication.

14. The parties shall comply with any request or direction of the adjudicator in relation to the adjudication.

15. If, without showing sufficient cause, a party fails to comply with any request, direction or timetable of the adjudicator made in accordance with his powers, fails to produce any document or written statement requested by the adjudicator, or in any other way fails to comply with a requirement under these provisions relating to the adjudication, the adjudicator may–

(a) continue the adjudication in the absence of that party or of the document or written statement requested,

(b) draw such inferences from that failure to comply as circumstances may, in the adjudicator's opinion, be justified, and

(c) make a decision on the basis of the information before him attaching such weight as he thinks fit to any evidence submitted to him outside any period he may have requested or directed.

16.–(1) Subject to any agreement between the parties to the contrary, and to the terms of paragraph (2) below, any party to the dispute may be assisted by, or represented by, such advisers or representatives (whether legally qualified or not) as he considers appropriate.

(2) Where the adjudicator is considering oral evidence or representations, a party to the dispute may not be represented by more than one person, unless the adjudicator gives directions to the contrary.

17. The adjudicator shall consider any relevant information submitted to him by any of the parties to the dispute and shall make available to them any information to be taken into account in reaching his decision.

18. The adjudicator and any party to the dispute shall not disclose to any other person any information or document provided to him in connection with the adjudication which the party supplying it has indicated is to be treated as confidential, except to the extent that it is necessary for the purposes of, or in connection with, the adjudication.

19.–(1) The adjudicator shall reach his decision not later than–

(a) twenty eight days after the date of the referral notice mentioned in paragraph 7(1), or

(b) forty two days after the date of the referral notice if the referring party so consents, or
(c) such period exceeding twenty eight days after the referral notice as the parties to the dispute may, after the giving of that notice, agree.

(2) Where the adjudicator fails, for any reason, to reach his decision in accordance with paragraph (1)
  (a) any of the parties to the dispute may serve a fresh notice under paragraph 1 and shall request an adjudicator to act in accordance with paragraphs 2 to 7; and
  (b) if requested by the new adjudicator and insofar as it is reasonably practicable, the parties shall supply him with copies of all documents which they had made available to the previous adjudicator.

(3) As soon as possible after he has reached a decision, the adjudicator shall deliver a copy of that decision to each of the parties to the contract.

Adjudicator's decision

20. The adjudicator shall decide the matters in dispute. He may take into account any other matters which the parties to the dispute agree should be within the scope of the adjudication or which are matters under the contract which he considers are necessarily connected with the dispute. In particular, he may–

  (a) open up, revise and review any decision taken or any certificate given by any person referred to in the contract unless the contract states that the decision or certificate is final and conclusive,
  (b) decide that any of the parties to the dispute is liable to make a payment under the contract (whether in sterling or some other currency) and, subject to section 111(4) of the Act, when that payment is due and the final date for payment,
  (c) having regard to any term of the contract relating to the payment of interest decide the circumstances in which, and the rates at which, and the periods for which simple or compound rates of interest shall be paid.

21. In the absence of any directions by the adjudicator relating to the time for performance of his decision, the parties shall be required to comply with any decision of the adjudicator immediately on delivery of the decision to the parties in accordance with this paragraph.

22. If requested by one of the parties to the dispute, the adjudicator shall provide reasons for his decision.

Effects of the decision

23.-(1) In his decision, the adjudicator may, if he thinks fit, order any of the parties to comply peremptorily with his decision or any part of it.

(2) The decision of the adjudicator shall be binding on the parties, and they shall comply with it until the dispute is finally determined by legal proceedings, by arbitration (if the contract provides for arbitration or the parties otherwise agree to arbitration) or by agreement between the parties.

24. Section 42 of the Arbitration Act 1996 shall apply to this Scheme subject to the following modifications–
  (a) in subsection (2) for the word "tribunal" wherever it appears there shall be substituted the word "adjudicator",
  (b) in subparagraph (b) of subsection (2) for the words "arbitral proceedings" there shall be substituted the word "adjudication",
  (c) subparagraph (c) of subsection (2) shall be deleted, and
  (d) subsection (3) shall be deleted.

25. The adjudicator shall be entitled to the payment of such reasonable amount as he may determine by way of fees and expenses reasonably incurred by him. The parties shall be jointly and severally liable for any sum which remains outstanding following the making of any determination on how the payment shall be apportioned.

26. The adjudicator shall not be liable for anything done or omitted in the discharge or purported discharge of his functions as adjudicator unless the act or omission is in bad faith, and any employee or agent of the adjudicator shall be similarly protected from liability.

PART II–PAYMENT

Entitlement to and amount of stage payments

1. Where the parties to a relevant construction contract fail to agree–
  (a) the amount of any instalment or stage or periodic payment for any work under the contract, or
  (b) the intervals at which, or circumstances in which, such payments become due under that contract, or
  (c) both of the matters mentioned in sub-paragraphs (a) and (b) above,
the relevant provisions of paragraphs 2 to 4 below shall apply.

2.-(1) The amount of any payment by way of instalments or stage or periodic payments in respect of a relevant period shall be the difference between the amount determined in accordance with sub-paragraph (2) and the amount determined in accordance with sub-paragraph (3).

(2) The aggregate of the following amounts–
  (a) an amount equal to the value of any work performed in accordance with the relevant construction contract during the period from the commencement of the contract to the end of the relevant period (excluding any amount calculated in accordance with sub-paragraph (b)),
  (b) where the contract provides for payment for materials, an amount equal to the value of any materials manufactured on site or brought onto site for the purposes of the works during the period from the commencement of the contract to the end of the relevant period, and
  (c) any other amount or sum which the contract specifies shall be payable during or in respect of the period from the commencement of the contract to the end of the relevant period.

(3) The aggregate of any sums which have been paid or are due for payment by way of instalments, stage or periodic payments during the period from the commencement of the contract to the end of the relevant period.

(4) An amount calculated in accordance with this paragraph shall not exceed the difference between–
  (a) the contract price, and
  (b) the aggregate of the instalments or stage or periodic payments which have become due.

Dates for payment

3. Where the parties to a construction contract fail to provide an adequate mechanism for determining either what payments become due under the contract, or when they become due for payment, or both, the relevant provisions of paragraphs 4 to 7 shall apply.

4. Any payment of a kind mentioned in paragraph 2 above shall become due on whichever of the following dates occurs later–
  (a) the expiry of 7 days following the relevant period mentioned in paragraph 2(1) above, or
  (b) the making of a claim by the payee.

5. The final payment payable under a relevant construction contract, namely the payment of an amount equal to the difference (if any) between–
  (a) the contract price, and
  (b) the aggregate of any instalment or stage or periodic payments which have become due under the contract,
shall become due on the expiry of–

(a) 30 days following completion of the work, or
(b) the making of a claim by the payee,
whichever is the later.

6. Payment of the contract price under a construction contract (not being a relevant construction contract) shall become due on
(a) the expiry of 30 days following the completion of the work, or
(b) the making of a claim by the payee,
whichever is the later.

7. Any other payment under a construction contract shall become due
(a) on the expiry of 7 days following the completion of the work to which the payment relates, or
(b) the making of a claim by the payee,
whichever is the later.

Final date for payment

8.-(1) Where the parties to a construction contract fail to provide a final date for payment in relation to any sum which becomes due under a construction contract, the provisions of this paragraph shall apply.

(2) The final date for the making of any payment of a kind mentioned in paragraphs 2, 5, 6 or 7, shall be 17 days from the date that payment becomes due.

Notice specifying amount of payment

9. A party to a construction contract shall, not later than 5 days after the date on which any payment–
(a) becomes due from him, or
(b) would have become due, if–
   (i) the other party had carried out his obligations under the contract, and
   (ii) no set-off or abatement was permitted by reference to any sum claimed to be due under one or more other contracts,
give notice to the other party to the contract specifying the amount (if any) of the payment he has made or proposes to make, specifying to what the payment relates and the basis on which that amount is calculated.

Notice of intention to withhold payment

10. Any notice of intention to withhold payment mentioned in section 111 of the Act shall be given not later than the prescribed period, which is to say not later than 7 days before the final date for payment determined either in accordance with the construction contract, or

where no such provision is made in the contract, in accordance with paragraph 8 above.

Prohibition of conditional payment provisions

11. Where a provision making payment under a construction contract conditional on the payer receiving payment from a third person is ineffective as mentioned in section 113 of the Act, and the parties have not agreed other terms for payment, the relevant provisions of–
- (a) paragraphs 2, 4, 5, 7, 8, 9 and 10 shall apply in the case of a relevant construction contract, and
- (b) paragraphs 6, 7, 8, 9 and 10 shall apply in the case of any other construction contract.

Interpretation

12. In this Part of the Scheme for Construction Contracts–

"claim by the payee" means a written notice given by the party carrying out work under a construction contract to the other party specifying the amount of any payment or payments which he considers to be due and the basis on which it is, or they are calculated;

"contract price" means the entire sum payable under the construction contract in respect of the work;

"relevant construction contract" means any construction contract other than one–
- (a) which specifies that the duration of the work is to be less than 45 days, or
- (b) in respect of which the parties agree that the duration of the work is estimated to be less than 45 days;

"relevant period" means a period which is specified in, or is calculated by reference to the construction contract or where no such period is so specified or is so calculable, a period of 28 days;

"value of work" means an amount determined in accordance with the construction contract under which the work is performed or where the contract contains no such provision, the cost of any work performed in accordance with that contract together with an amount equal to any overhead or profit included in the contract price;

"work" means any of the work or services mentioned in section 104 of the Act.

EXPLANATORY NOTE

(This note is not part of the Order)

Part II of the Housing Grants, Construction and Regeneration Act 1996 makes provision in relation to construction contracts. Section 114 empowers the Secretary of State to make the Scheme for Construction

Contracts. Where a construction contract does not comply with the requirements of sections 108 to 111 (adjudication of disputes and payment provisions), and section 113 (prohibition of conditional payment provisions), the relevant provisions of the Scheme for Construction Contracts have effect.

The Scheme which is contained in the Schedule to these Regulations is in two parts. Part I provides for the selection and appointment of an adjudicator, gives powers to the adjudicator to gather and consider information, and makes provisions in respect of his decisions. Part II makes provision with respect to payments under a construction contract where either the contract fails to make provision or the parties fail to agree–
- (a) the method for calculating the amount of any instalment, stage or periodic payment,
- (b) the due date and the final date for payments to be made, and
- (c) prescribes the period within which a notice of intention to withhold payment must be given.

Table of Cases

ABB Power Construction Ltd *v.* Norwest Holst Engineering Ltd
[2000] BLR 426 .... 21, 22
A&D Maintenance and Construction Ltd *v.* Pagehurst Construction
Services Ltd 17-CLD-09-07; (1999) CILL 1518 ...... 44, 47, 194
Absolute Rentals Ltd *v.* Gencor Enterprises Ltd (2000) CILL 1637
.... 34, 189, 195
Allied London and Scottish Properties plc *v.* Riverbrae Construction
Ltd 17-CLD-08-01; [1999] BLR 46 .... 139, 192
Ashville Investments *v.* Elmer Contractors Ltd [1989] QB 488 ... 43
Atlas Ceiling & Partition Ltd (The) *v.* Crowngate Estates
(Cheltenham) Ltd (2000) CILL 1639 .... 32, 33, 95

Beaufort Developments (NI) Ltd *v.* Gilbert Ash NI Ltd (1998)
88 BLR 1 134
Bloor Construction (UK) Ltd *v.* Bowmer & Kirkland (London) Ltd
(2000) CILL 1626 153
Bouygues UK Ltd *v.* Dahl-Jensen UK Ltd (2000) 17-CLD-06-11;
(CA) TLR 17 August 2000
.... 9, 49, 98, 100, 103, 146, 151, 152, 169, 190, 193, 196
Bridgeway Construction Ltd *v.* Tolent Construction Ltd (2000)
www.adjudication.co.uk/cases.htm 55
British Steel Corporation *v.* Cleveland Bridge Engineering Co Ltd
[1984] 1 All ER 504; (1983) BLR 94 33
Brown *v.* Llandovery Terra Cotta etc. Co Ltd (1909) 25 TLR 625
.... 165

Carter (R.G.) Ltd *v.* Edmund Nuttall Ltd (2000)
www.adjudication.co.uk/cases.htm 37, 42, 60
Christiani & Nielsen Ltd v. The Lowry Centre Development Co Ltd
(2000) www.adjudication.co.uk/cases/christiani.htm
.... 31, 100, 194
Cook (F.W.) Ltd *v.* Shimizu (UK) Ltd (2000) CILL 1613
.... 180-181, 194
Re Cooms and Freshfield and Fernley (1850) 4 Exch 839;
154 ER 1456 166

Cornhill Insurance plc v. Improvement Services Ltd (1986)
2 BCC 98 182
Costain Building and Civil Engineering Ltd v. Scottish Rugby Union
plc [1993] SC 650 216
Crampton and Holt v. Ridley & Co (1887) 20 QBD 48 165
Cruden Construction Ltd v. Commission for the New Towns (1995)
CILL 1035 39
Culling v. Tufnal (1694) Bull NP 34 18

Discain Project Services Ltd v. Opecprime Developments Ltd LTL
12/10/2000 .... 8, 103, 114, 201

Edmund Nuttall Ltd v. Sevenoaks District Council LTL 27/9/2000
.... 153
Elanay Contracts Ltd v. The Vestry LTL 13/10/2000 .... 196, 204

Fastrack Contractors Ltd v. Morrison Construction Ltd and
Impreglio UK Ltd 17-CLD-05-01; (2000) CILL 1589
.... 61, 98–99
Fillite (Runcorn) v. Aqua-Lift (1989) 45 BLR 27 43

Gilbert Ash (Northern) Ltd v. Modern Engineering (Bristol) Ltd
[1974] AC 689; [1973] 3 All ER 195 .... 4, 198–199
Gillespie Brothers & Co Ltd v. Roy Bowles Transport Ltd [1973]
1 QB 400 155
Grovedeck Ltd v. Capital Demolition Ltd 17-CLD-03-10;
[2000] BLR 181 .... 36, 45, 193

Halki Shipping Corporation v. Sopex Oils Ltd [1998] 1 WLR 726
.... 40, 188
Hanak v. Green [1958] 2 QB 9; [1958] 2 All ER 141 ..... 198, 199
Hatzfeld-Wildenburg v. Alexander [1912] 1 Ch 248 32
Hayter v. Nelson and Home Insurance Co [1990] 2 Lloyds Rep 265
.... 40, 186
Herschel Engineering Ltd v. Breen Property Ltd (2000) CILL 1616
.... 60, 139, 195
Heyman v. Darwins [1942] AC 356; [1942] All ER 337 ..... 43, 92
Hobday v. Nash [1944] 1 All ER 302 17
Homer Burgess Ltd v. Chirex (Annan) Ltd 17-CLD-06-01; (2000)
CILL 1580 .... 21, 97, 98, 192, 193

John Barker Construction Ltd v. London Portman Hotel Ltd (1996)
83 BLR 31 135

John Cothliff Ltd v. Allen Build (North West) Ltd 17-CLD-09-04;
   (1999) CILL 1530 169
John Mowlem plc v. Hydra-Tight Ltd (2000) CILL 1650 ... 42, 59
Jones v. Sherwood Computer Services plc [1992] 1 WLR 277 ... 9

K&D Contractors v. Midas Homes Ltd (2000)
   www.adjudication.co.uk/cases.htm 62

Laker Airways Inc v. FLS Aerospace Ltd 17-CLD-07-28; (1999)
   CILL 1508 102
Lathom Construction Ltd v. Cross and Cross (2000) CILL 1568
   .... 44, 193
Lubenham Fidelities and Investment Co v. South Pembrokeshire
   District Council (1986) 33 BLR 39 216

Macob Civil Engineering Ltd v. Morrison Construction Ltd [1999]
   BLR 93 ....
   9, 113, 122, 134, 142, 151, 170, 176, 178, 179, 182, 183, 188, 201, 208
Melton Medes Ltd & Another v. Securities and Investment Board
   [1995] 3 All ER 881 .... 55, 157
Mills & Rockleys Ltd v. Leicester City Council [1946] KB 315,
   [1946] 1 All ER 424 17
Minshall v. Lloyd (1837) 2 M&W 450; 150 ER 705 18
Modern Engineering (Bristol) Ltd v. Gilbert Ash (Northern) Ltd see
   Gilbert Ash (Northern) Ltd v. Modern Engineering (Bristol) Ltd
Monk Construction Ltd v. Norwich Union (1992) 62 BLR 107
   33

Nikko Hotels (UK) Ltd v. MEPC plc [1991] 2 EGLR 103 ... 9, 191
Nottingham Community Housing Association v. Powerminster Ltd
   (2000) www.adjudication.co.uk/cases/powerminster.htm
   18
Northern Developments (Cumbria) Ltd v. J&J Nichol
   17-CLD-05-19; [2000] BLR 158
   .... 43, 91, 170-171, 194, 200-201, 228-229
Northern Regional Health Authority v. Derek Crouch Construction
   Co [1984] QB 644; 2 All ER 175 134

Outwing Construction Ltd v. H Randell and Son Ltd (1999)
   89 BLR 156 .... 179, 184

Palmers Ltd v. ABB Power Construction Ltd 17-CLD-07-01; (1999)
   CILL 1543 .... 18, 19, 22, 95, 100, 193

Pepper v. Hart [1993] AC 593 139
Project Consultancy Group (The) v. The Trustees of the Gray Trust
    17-CLD-09-01; [1999] BLR 377 .... 31, 94, 97, 192-193, 194

R v. Bow Street Magistrate, ex parte Pinochet (No 2) [1999]
    1 All ER 577 102
R v. Cripps [1984] QB 686 154
R v. Gough [1993] AC 646 102
Roberts v. Eberhardt (1858) 3 CBNS 482; 140 ER 629 166

Shepherd Construction Ltd v. Mecright Ltd (2000)
    www.adjudication.co.uk/cases.htm 44, 193
Sherwood & Casson Ltd v. Mackenzie Engineering Ltd
    17-CLD-03-27; (2000) CILL 1577 .... 80, 97, 98 194
Stein v. Blake [1996] AC 243 .... 197, 198
Straume (A.) (UK) Ltd v. Bradlor Developments Ltd (1999)
    CILL 1518 .... 48, 201

Re Taylor's Industrial Flooring Ltd [1990] BCC 44 182
Tim Butler Contractors Ltd v. Merewood Homes Ltd (2000)
    www.adjudication.co.uk/cases/butler.htm 191, 192
Trollope & Colls Ltd v. Atomic Power Constructions Ltd [1963]
    1 WLR 1035 32
Turriff Construction Ltd v. Regalia Knitting Mills Ltd (1971)
    9 BLR 20 33

VHE Construction PLC v. RBSTB Trust Co Ltd 17-CLD-05-09; (2000)
    CILL 1592 200

Westminster Chemicals & Produce Ltd v. Eicholz & Loeser [1954]
    1 Lloyds Rep 99 95
Whiteways Contractors (Sussex) Ltd v. Impresa Castelli
    Construction UK Ltd LTL 23/8/2000 230
Workplace Technologies v. E. Squared Ltd and Mr J.L. Riches (2000)
    17-CLD-02-23 96

References to Housing Grants, Construction and Regeneration Act 1996

Section	Page	Section	Page
104(1)	23–4, 205	109(1)	206, 220
104(2)	23–4	109(2)	206
104(3)	26–7	109(3)	206
104(5)	30	109(4)	206
104(6)	30–31, 33	110(1)	113, 208, 209, 212, 230
104(7)	33	110(2)	207, 209–10, 226
105(1)	16–19, 22	110(3)	208
105(2)	19–22, 97	111(1)	199, 210–11, 227, 228, 230
106(1)	34, 205	111(2)	199, 211
106(2)	34, 205	111(3)	199–200, 211
107	35–8	111(4)	127, 137, 200, 211
108(1)	39, 43, 55, 56, 57	114(4)	56
108(2)	8, 46–52, 55, 56, 57, 59, 83, 101, 105, 120	113	215–18, 230
108(3)	46, 53, 55, 56, 61, 94, 150, 174, 203	112	212–14
		115	66
108(4)	46, 54–5, 56, 154	116	51, 121

References to Scheme for Construction Contracts

Part I Adjudication

Paragraph	Page
1(1)	57
1(2)	66–7
1(3)	60–61, 62, 129
2(1)	68–9, 70
2(2)	73
2(3)	69
4	67, 75, 101–2
5	71, 74
6(1)	73–4
7(1)	84, 86
7(2)	86
7(3)	86
8	61
9(1)	79, 127, 160
9(2)	79, 80–81, 98
9(3)	79, 80
9(4)	80
10	75
11	76–8, 123, 159, 160
12	68, 76, 78, 101, 102, 113–14
13	105–9, 115, 128
14	112–13
15	111, 114
16	116–18
17	78, 108–11, 118, 128
18	118
19	78
19(1)	120–21
19(2)	121–3
19(3)	123, 166
20	91–2, 101, 127, 135, 136, 140, 142, 144
21	130, 144–5
22	78, 145
23	61, 145, 150, 151, 174
24	145, 174–7
25	77, 102, 158, 159, 160
26	154–5

Part II Payment

Paragraph	Page
1	219
2	219–20
3	222
4	222, 223–4
5	222–3, 224
6	223
7	223
8	225–6, 228
9	226
10	228
11	231
12	220, 221, 224

Subject Index

action, limitation of, 48-9
adjudication
  arbitration compared with, 8-9, 53-4
  conduct of, 101
  costs of, 55-6
  definition of, 6-10
  expert determination, compared with, 14
  future of, 13-15
  litigation compared with, 7-8
  mediation compared with, 10-11
  notice of, 57-8, 60, 129-30
    service, 66-7
  right to refer, 39, 46-50
  timetable for, 50-53, 57-60, 83, 101, 115
  under JCT 81, 5
  under subcontracts prior to the Act, 4
adjudicator
  bad faith of, 157
  costs, power to order payment of, 169
  death of, 81
  discretion re procedures, 103
  duties of, 101
  duty to act impartially, 52, 101-102
  duty to decide, 101, 127-8
  failure of appointment, 73-5
  fees and expenses, 77, 158
    security, 123, 165
  immunity of, 54-5, 154
  initiative, ability to take, 52-3, 105
  jurisdiction of *see* jurisdiction
  liability of, 54-5
  misconduct of, 78
  objection to, 75-6
  preliminary directions of, 106
  qualifications, 12-13, 67
  resignation, 76, 98, 122, 127
  revocation of appointment, 76
  selection of, 67
  terms of appointment, 71
  third parties, duty towards, 156
adjudicator nominating bodies, 12-13, 51, 58, 65, 69, 76
  request to appoint, 70-71
adjudicators
  errors by, 15
  fees of
    consistency in, 15
    security for, 15
  insurance required by, 157
  numbers of, 12-13
administration, company in, 48
appeal, 11-12
arbitration
  adjudication, compared with, 8-9, 42
  right to refer disputes to, 40, 185-90
  to enforce decision, 183-5

bad faith, 54-5

Centre for Dispute Resolution (CEDR), 71, 132, 136, 148, 155, 164, 167, 172
certificates, power to open up, 134
Chartered Institute of Arbitrators, 13, 70

260 Subject Index

Civil Engineering Contractors
    Association (CECA) contract,
    59
cleaning of buildings, 18–19
collateral warranties, 25
commencement date, 30-33
compliance with adjudicator's
    directions, failure to comply,
    112
compromise agreements, 44
confidentiality, 118–20
Constructing the Team report, 1
construction contract, 16
    definition of, 22–5
    exclusions from, 26–30
Construction Contracts (England and
    Wales) Exclusion Order 1998,
    26
Construction (Design and
    Management) Regulations
    1994, 25
Construction Industry Council (CIC),
    12
    Model Adjudication Procedure and
        Agreement, 54, 72, 81, 85–6, 93,
        112, 119, 124, 133, 136, 148, 150,
        155, 156, 161–2, 164, 167
construction management, 23
construction operations
    definition of, 16–19
    exclusions from definition, 19–22,
        95–6
    outside United Kingdom, 33
consultancy contracts, 24
contract
    compromise, 44
    date of, 31–2
    determination of, 47
    repudiation of, 43–4, 194
costs, 15, 145, 158
counterclaims, 130
court proceedings, concurrent, 60,
    195

decision of adjudicator
    binding, 53–4, 150
    enforcement of, 174
        arbitration, 183–5
        challenges to, 185
        injunction, 177, 182
        peremptory orders, 174–7
        summary judgment, 178–81
        winding up petition, 182–3
    form and content of, 142
    mistakes in, 152–4, 190–92
    reasons for, 145
        failure to give, 79
        form of, 149
    timetable for, 120
deed, date of execution, 32
development agreements, 29
difference, meaning of, 39–42
Disputes
    definition of, 39–42, 57
    multiple, 15
    related, 91–3
    under several contracts, 45
    under the contract, 43, 91

employment, contract of, 26
Engineering and Construction
    Contract *see* New Engineering
    Contract
expert determination, adjudication
    compared with, 9–10
expert, independent, 109, 112, 128–9,
    150

facilities management, 18
finance agreements, 28–9

Glasgow Caledonian University, 12
Government contract GC/Wks/1, 59,
    69, 72, 86, 120, 123, 124, 132,
    147, 152, 163, 167

Hansard, 37

heating systems, maintenance of, 18
Housing Grants, Construction and Regeneration Bill, 3
Human Rights Act, 14, 202–4

ICE Adjudication Procedure, 72, 85, 104–5, 124, 126, 133, 136, 148, 153, 162–3, 167, 172
ICE Standard Form of Contract, clause 66, 50, 59
insolvency, 48, 139, 195–8, 217
insurers, 14
interest, 15, 140–41, 144

JCT contracts, 59, 69, 71, 81, 104, 116, 120, 125–6, 132, 136, 142, 148, 151–2, 167, 172
   April 1998 amendments, 85, 124
JCT Standard Form of Adjudicator's Agreement, 54, 69, 71–2, 85, 107, 124, 125–6, 160–61
JCT Standard Form of Contract 98, condition 41A, 50
judicial review, 11–12
jurisdiction, 93, 134, 192–5
   adjudicator's power to determine, 98
   adjudicator should have resigned, 98, 194–5
   concurrent court proceedings, 195
   contract not for construction operations, 95–6, 97, 193
   contract not in writing, 193
   contract pre 1 May 1998, 94–5, 194
   dispute not covered by adjudication agreement, 194
   dispute not covered by notice of adjudication, 194

labour only contracts, 24
language, 108
Latham, Sir Michael, 1–3, 6
letter of intent, 32–3

limitation of action, 48–9
litigation, adjudication compared with, 7–8

maintenance, 18
making the Scheme for Construction Contracts, 4
management contract, 23
mediation, adjudication compared with, 10–11
mixed contracts, 30

National Health Service (Private Finance) Act 1997, 27
natural justice, 8, 14, 103, 113, 201–2
New Engineering Contract, 2, 6, 59, 148, 167
novation, 25

Official Referees Court *see* Technology and Construction Court

painting and decorating, 19
payment, 205
   amount, notice of, 209–10, 226–8
   conditional, 214–17, 230–31
   decisions requiring, 136, 144
   final date for, 137, 144, 225–6
   notice of intention to withhold, 210–12, 228–30
   right to stage payments, 206–7, 219–22
   time for, 144
   timing and quantification of, 208–9, 220
pay when paid, 214–17, 230–31
peremptory order, 174–6
Planning Supervisor, Standard Form of Appointment of, 25
Private Finance Initiative, 27

referral notice, 57, 81, 83, 107
representation, 116–18

residential occupiers, 34–5
Royal Institution of Chartered
    Surveyors, 13, 70
  arbitrators panel, 96

scaffolding, 19, 22
Scheme for Construction Contracts,
    draft, 4
Scheme for Construction Contracts
    (England and Wales)
    Regulations 1988 (the Scheme),
    4
  amendment of, 56
security agreements, 29
set-off, 198–201
Speaight, Anthony QC, 11
stakeholders account, 56, 138–9
structure, definition of, 17
suspend, right to, 212–14

Technology and Construction Court,
    2, 10, 14
Technology and Construction
    Solicitors Association, 13, 71, 81
  adjudication rules, 60, 86, 93, 104,
    111–12, 118, 124, 133, 136, 142,
    148, 155, 163–4, 167, 172
third parties, duty of adjudicator
    towards, 156
trust fund, 3

Value Added Tax, 144, 161

Wolverhampton, University of, 13
Woolf, Lord, 2
Woolf Report, 2
writing
  contract evidenced in, 35–8
  contract in, 35–8, 193